Springer Geography

The Springer Geography series seeks to publish a broad portfolio of scientific books, aiming at researchers, students, and everyone interested in geographical research. The series includes peer-reviewed monographs, edited volumes, textbooks, and conference proceedings. It covers the entire research area of geography including, but not limited to, Economic Geography, Physical Geography, Quantitative Geography, and Regional/Urban Planning.

More information about this series at http://www.springer.com/series/10180

Waldemar Cudny

Festivalisation of Urban Spaces

Factors, Processes and Effects

Springer

Waldemar Cudny
Lodz University
Lodz
Poland

ISSN 2194-315X ISSN 2194-3168 (electronic)
Springer Geography
ISBN 978-3-319-31995-7 ISBN 978-3-319-31997-1 (eBook)
DOI 10.1007/978-3-319-31997-1

Library of Congress Control Number: 2016940005

Printed on acid-free paper

This Springer imprint is published by Springer Nature
The registered company is Springer International Publishing AG Switzerland

For my mother

Preface

The book is a multifaceted monograph on festivals and their role in shaping the multidimensional geographical urban space. It focuses on the issue of urban festivals, their types and impacts on urban space in the process of festivalisation. The intention was to create a book which will lead the reader all the way from the analysis of the basic concepts related to festivals and festivalisation, through the history of festivals and their development factors to the different impacts they have on space and the methods of studying festivals. The discussion is illustrated with numerous practical examples, including a detailed case study presented in one of the chapters.

The book explores the problems of festivalisation for several reasons. Firstly, its author comes from Poland—a country which regained true independence a relatively short ago, in 1989, after 40 years of the communist regime. One result of the rapid change of the political and economic system in Poland in the 1990s was an intensive development of festivals.

At present, events play a significant role in shaping the multidimensional urban space of many Polish cities. Although festivals existed before 1989 as well, their number, scale, variety and impacts were not so large as nowadays. The book is written by an urban geographer by education and by passion. He noticed the growing role of culture, particularly festivals, in the development of post-socialist cities, while conducting his research in Poland. Events of this type are playing an increasing role in shaping and developing multidimensional urban space. These observations were an impulse to undertake further theoretical and practical studies. The theoretical analysis led to the conclusion that the festivalisation phenomena occurring in Poland in recent decades are strongly connected with similar tendencies observed earlier in Western countries. Literature analysis undertaken for the purposes of this book allowed to specify conceptual and methodological issues as well as the most important groups of impacts exerted by festivals on urban space. Many observations described in this publication were confirmed in practical research conducted in the form of case studies, e.g. in the area of Łódź (Poland).

Secondly, the author has been concerned with the problems of events and festivals for many years. Thus, it was decided to resent this research experience in the form of a scientific monograph. Waldemar Cudny wrote his first work regarding festivalisation in 2006, concerning the Dialogue of Four Cultures Festival in Łódź (Poland). Since then, he has published (as the author and co-author) several works concerning events and festivals. They were both theoretical and practical works, based on the author's own field studies and observations. The works present the main problems related to festivals, festivalisation and festival tourism, the types of festivals and their influence on geographical space. Analyses based on field studies included mostly case studies of various festivals organised in Łódź—a Polish post-industrial city. Waldemar Cudny is also a specialist in the field of broadly understood socio-economic and spatial transformations taking place in cities, especially post-socialist cities. His research in the field of the socio-economic analysis of cities was published in the form of several dozen publications, including articles, chapters in monographs as well as whole monographs on urban geography. This solid foundation in the form of own theoretical and practical studies of festivalisation and the transformations in cities enabled him to collect sufficient material and gain enough knowledge to write this book.

Another inspiration to write this book was the research gap concerning the festivalisation process. There are no publications on the book market which would present the phenomenon of festivals in a comprehensive way, starting from their definition, through history to the description of their influence. There is also a lack of more advanced books showing how festivals are used for the transformation and development of urban space, i.e. books written from the perspective of urban geography or, more broadly, urban studies. The intention of the book was to create a publication which would fill this research gap, at least in part.

As regards the potential readers of the book, it is intended for a large and varied group. Multidisciplinary character of the publication is its major advantage, which significantly broadens the range of readers. The book draws on many sciences: event studies, cultural and tourism geography, sociology, anthropology, etc. It is also based on the analysis of issues typical of urban geography, urban economy or urban development planning, as well as territorial marketing. Furthermore, on the one hand, the book has been thought as a publication presenting human geography issues, while on the other, it shows the theoretical and practical factors and processes of urban development. As regards the academic milieu, the intention was to construct the book in such a way that it could be used by specialists concerned with broadly understood culture, researchers as well as students interested in urban problems. Therefore, the publication will be useful for representatives of culture studies, anthropology and cultural geography, on the one hand, and researchers concerned with urban geography, urban economy, urban development planning or territorial marketing on the other. Events, including the festivals described in this book, are successfully used nowadays for shaping urban space, generating economic phenomena and creating a positive image of cities and urban marketing.

Researchers will find here a number of interesting ideas, including theoretical issues, such as research themes, concepts or the impacts of festivals on space. Students are an important group of readers as well. The book will provide them with the knowledge of the basic concepts and approaches to the phenomenon of festivals. They will also find here the methodological foundations of the study of festivals and festivalisation. Separate part of this work is devoted to the description of issues connected with the methods of researching festivals. This part of the publication will be particularly useful to students, because it will help them choose the proper methods to study the phenomenon on their own. Information of this type will certainly be helpful in choosing the topics of and writing their diploma theses (BA or MA) as a part of culture, urban geography and urban economy studies. The publication will also serve as a major literature source for students studying related subjects.

The book is constructed in such a way that it covers a range of theoretical and practical issues. It presents a discussion based on literature analysis, as well as many practical examples concerning, for instance, the history of festivals, their types or finally their impacts on urban space. On the other hand, the work contains a vast chapter dedicated to the process of festivalisation in a Polish post-socialist city of Łódź. The chapter presents a large case study concerning the location, history of the city and the changes of its cultural function after 1989. The author also conducted a wide-ranging analysis of the main festivals, which were organised in the city after 1989. Apart from that, the study concerned the ways in which those festivals participated in the transformation of the multidimensional space of a post-socialist and post-industrial city, such as Łódź. Moreover, the work contains a model of the influence of festivalisation on urban space presented from the social, economic, spatial and marketing perspective. It was based on the theoretical and practical analysis, the results of which were presented in the previous parts of the book. The intention was to make the model universal enough to be used for further comparative analysis as regards other cities in the world.

The book has been written for a large group of readers, including researchers, students and practitioners concerned with culture and cities. In this respect, the publication differs from other books describing festivals and other events. There are quite many such publications on the market, but they have a different character than this one. They usually present terminology, practical aspects of organising events, their influence on tourist traffic or finally classifications of events. Books presenting methods of investigating festivals are rarer, and there are really few books which give a comprehensive picture of festivals and their impacts on urban space. Combining various study approaches, borrowed from, e.g. event studies, culture studies, sociology or urban studies, the book will fill in the research gap on the international publishing market and will become a solid base for further comparative analysis by theoreticians, students and practitioners concerned with urban festivals.

Contents

About the Author

Dr. Waldemar Cudny gained his education in Poland, at the Faculty of Geographical Sciences, University of Łódź. He received the Ph.D. degree in earth sciences, in the field of socio-economic geography in 2004 and the MA degree in geography, in the field of spatial economy and spatial urban planning in 1999. Since 2005, he has been employed as an assistant professor at the Institute of Tourism and Economic Development, in the Tomaszów Mazowiecki branch of the University of Łódź. Dr. Waldemar Cudny specialises in two main research themes. The first of them is the analysis of the socio-economic and spatial transformations taking place in European post-socialist cities, and the other one are issues related to cultural geography as well as cultural and festival tourism. By 2015, Dr. Waldemar Cudny has published (individually and as a co-author) the total of over 50 reviewed scientific works in the form of books, articles and chapters in monographs. In 2006, he published his first scientific monograph, devoted to the socio-economic and spatial changes of the town of Tomaszów Mazowiecki during the post-socialist system transformation. In 2012, as a co-author and co-editor, he published another scientific monograph entitled *Tourism and the Transformation of Large Cities in the Post-Communist Countries of Central and Eastern Europe*, comparing the development of urban tourism in five post-socialist countries of Central and Eastern Europe. Dr. Waldemar Cudny is also an experienced academic teacher. In his career so far, he has taught subjects related to human geography, urban and cultural geography and tourism. In recent years, he has done research and conducted classes at foreign universities, e.g. the Comenius University in Bratislava (Slovakia), the Catholic University of Eichstätt-Ingolstadt and Justus Liebig-Universität in Giessen (Germany), Umea University (Sweden) and Birmingham University (Great Britain).

Author's Selected Scientific Publications:

Cudny W (2014) Festivals as a subject for geographical research. Geografisk Tidsskrift-Danish Journal of Geography 114(2):132–142

Cudny W (2014) The Phenomenon of Festivals—Their Origins, Evolution and Classifications, Anthropos 109(2):640–656

Cudny W (2014) The influence of the „Komisarz Alex" TV series on the development of Łódź in the eyes of the city inhabitants. Moravian Geographical Reports 22(1):33–43

Cudny W (2013) Festival tourism—the concept, key functions and dysfunctions in the context of tourism geography studies. Geografický časopis 6(2):105–118

Cudny W, Rouba R (2013) A Large Sports Events in the Eyes of the Host City's Inhabitants: the Example of Wrocław. Polish Sociological Review 4(184):505–530

Cudny W (2012) Socio—Economic Changes in Lodz—Results of Twenty Years of System Transformation. Geografický časopis 64(1) 3–27

Cudny W (2012) Socio—economic transformation of small towns in East Germany after 1990—Colditz case study. Bulletin of Geography. Socio Economic Series 17:33–45

Cudny W, Michalski T, Rouba R (eds.) (2012) Tourism and the transformation of large cities in the post—communist countries of Central and Eastern Europe. ŁTN–Wydawnictwo Uniwersytetu Łódzkiego, Łódź

udny W (2006) Przemiany przestrzenne i funkcjonalne Tomaszowa Mazowieckiegow okresie transformacji. Szlakami Nauki nr 35, ŁTN, Łódź

Chapter 1
Introduction

Abstract The Introduction is the first factual chapter of this book, which signals its main theoretical assumptions and contains basic information concerning the object of study discussed in this book. Further on, it is shown how festivals and festivalisation relate to the socio-economic changes taking place in modern societies. This chapter also presents the correlation between festivalisation and the new socio-economic processes observed after the Second World War in the developed countries. Other problems signalled in the Introduction include the impact of the changes in post-communist countries on the development of festivalisation. Introduction also presents a research hypothesis which relates the development of festivalisation to the cultural turn, cultural economy, creative and experience societies, as well as the urban development processes that took place during the post-industrial period. Further on, the chapter presents a detailed list of the research objectives which are achieved in this book. The Introduction also contains a description of other basic issues, such as the research methods, or the structure of the work, which consists of the preface supplemented with a biographical note about the author and six factual chapters, followed by a summary. The Introduction ends with a brief description of the structure of the book, presenting the contents of individual chapters and subchapters.

Keywords Culture · Cultural turn · Culture and events · Introduction to festivals · Experience societies · Festivalisation · Urban development

Recent decades have brought a number of global civilisational (social and economic) changes. One of them was the growing interest in culture, which could be first observed in the developed countries of Europe and North America (so-called "cultural turn"). Interest in culture and its growing economic and social role started to have a significant influence on cities (Jacobs and Weiss Hanrahan 2005). Other important elements which have a positive effect on the development of culture and entertainment, e.g. in urban areas, are the play ethic and the formation of experience societies (see Schulze 1992; Rifkin 2000). Modern developed societies more and more often undertake activities connected with fun and interesting experiences.

W. Cudny, *Festivalisation of Urban Spaces*, Springer Geography,
DOI 10.1007/978-3-319-31997-1_1

A part of this phenomenon is the rapid development of cultural, entertainment or sports events. They are also very often used for restructuring and developing different urban spheres, e.g. social or economic, as well as for creating a positive image of cities (Smith 2012). In recent decades, changes strongly modifying contemporary urban space have included the processes of economy internationalisation and globalisation, which are nowadays significantly influencing urban space at many locations (see Amin 1994). We must not forget about the computer science revolution or the growing role of creativity and mobility in shaping contemporary urban communities, as well as the role of these phenomena in the development of events and event tourism (see Florida 2002; Urry 2007).

Moreover, in the 1990s, a large group of countries entered a new stage of historical and socio-economic development. The countries are former communist states, where the political system collapsed after the reforms of the Soviet *perestroika* and the Polish talks of the Round Table. After the fall of communism, cities in those countries underwent very rapid and complicated socio-economic and spatial transformations (see Słodczyk 2001; Sýkora and Bouzarovski 2012). They involved processes similar to those which had been taking place in capitalist western cities since the middle of the twentieth century (Ondoš and Korec 2008).

In the last decades, the developing cultural and entertainment activity has become a kind of a new urban development factor. An urban development factor is an element which contributes to the creation or further development of a city in a positive way (see Męczyński et al. 2010). Works referring to the analysis of the socio-economic development of cities present various typologies and descriptions of urban development factors. They can be divided into endogenous (internal) and exogenous (external); traditional and new; local, regional, national and international; direct and indirect; hard (easily measurable) and soft (difficult to measure); and controllable and incontrollable (see among others Parysek 1995, 1997; Wojtasiewicz 1997). The problems discussed in this book, i.e. culture, entertainment and the festival sector, should be currently regarded as new urban development factors. Classic factors are those elements which have an influence on the origins and the early development of cities. They include the land relief, access to water, the presence of mineral resources, accessibility by transport, population potential and state policy (earlier the ruler's politics). New factors include those elements which have been stimulating urban development for the last several decades: creativity, innovativeness, using culture for economic development (cultural economy), territorial marketing and city branding, as well as globalisation, which in some cities (e.g. the world largest metropolitan centres) is a powerful growth factor.

On the other hand, in some cases, festivals become a burden for the city. This happens in a situation when they are too costly, do not attract tourists, do not perform any other functions or cause problems, e.g. conflicts, overcrowding (as a result of the inflow of tourists) and higher prices. In this case, a festival which usually has a positive influence on an urban centre may hamper its development. The development of urban festivals and their impacts on the multidimensional urban space are of great importance nowadays. They are objects of interest in many sciences. They are also important from the practical point of view, e.g. to city planners or economists. Therefore, it was decided to consolidate Waldemar Cudny's research conducted in recent years in the form of this book, devoted to the broadly understood topic of urban festivals.

The object of study presented in this book is the phenomenon of festivals and their multidimensional influence on the geographical space in the city. Already Davies (2015, p. 533) pointed that "Many contemporary attempts to resolve the problems of the modernist city has been the relatively limited attention paid by most urban scholars to features of the city that relate to sensory not material attributes, to consumption not production, pleasure not work, and to episodic rather than permanent characteristics. Festive events in cities combine these four traits. These are activities consciously planed to last a limited time to mark special occasions, or to promote specific events, often using spectacles and parades designed to create excitement and joy, as well as achieving cultural, social and economic goals". Thus, this book is an attempt to create work that will fill in the research gap that regards the use of festivals in redevelopment of contemporary cities. This book investigates the basic issues related to festivals, i.e. the concepts, history, development factors and the problems arising in various dimensions of geographical urban space as a result of organising festivals. Furthermore, it presents the issues related to the theoretical side of festival analysis and the research methods, as well as some practical examples of the impacts of festival events exerted on cities. The publication is then a multifaceted monograph concerning urban festivals. It touches upon many research problems and themes explored in sciences concerned with culture and people, as well as in urban studies.

The problems of urban space festivalisation are studied in many scientific disciplines, such as urban geography, urban studies, cultural geography, event studies, culture science, anthropology or sociology. Clear references to these sciences are present in different parts this book. The issues discussed include the terminology and history of festivals. Analyses of this kind are based on human culture studies, which are a part of, e.g., anthropology. The influence of festivals on various spheres of human life, including the broad social or cultural sphere, is also studied in sociology or cultural geography. Festivals are commonly treated as a type of events, but they surely deserve a separate analysis, considering their unique character. Therefore, this publication represents works concerning festivals, written as a part of event studies. Another aspect of the problem discussed here is the influence of festivals on cities, with a particular consideration of urban space. In this case, the analysis was inspired by such scientific subdisciplines as urban geography and urban studies, which concern cities, networks of cities, their space and its tangible and intangible elements fulfilling the inhabitants' needs (see Pacione 2009). This publication can be regarded as interdisciplinary and multifaceted and therefore should be interesting to representatives of various scientific disciplines.

Moreover, the analysis presented in this publication is constructed in such a way that it can be used by both, theoreticians (e.g. researchers concerned with festival analysis) and practitioners who plan or organise events. The theoretical analysis is supported by many practical examples, e.g. of the types of impacts that festivals exert on urban space. In this way, this work would be helpful to people working in city halls or institutions concerned with culture or city promotion and organising or co-organising urban festivals. Apart from that, the publication has a didactic side too, as it encompasses a multifaceted analysis of festivals and festivalisation, starting from the definition of the phenomenon, through its history, functions and dysfunctions, research methods and a detailed case study (Chap. 5). From this point of view, the book may prove useful as a student's course book at university departments specialising in urban studies, urban geography, spatial planning, cultural geography, anthropology or sociology.

The main aims of the book include the following:

1. To present the concept of the festival, the analysis of the festival in contrast to other events. It is an important scientific issue as research on festivals should be unified across different sciences (see Getz 2010). Currently, there is considerable chaos in science as regards the concept of festivals; there are many definitions of the festival, and in different sciences, different approaches are taken to this problem. Therefore, the author decided to present his own understanding of the phenomenon, based on a review of various definitions. Besides, the aim of the book is to specify related concepts, such as festivalisation or festival tourism.
2. To present the perception of festivals in contemporary science and the key characteristics distinguishing them from other events, such as business (belonging to the MICE sector) or sports events, e.g. Football World or European Championships. Separating events of different types is another important research objective of this publication.
3. To present the history of festivals, which is an extremely interesting issue. It all started in antiquity and has lasted until the present day. It is important to show that festivals are not the invention of the twentieth century and that they were already popular in ancient Greece or in the Roman Empire. The idea is to present how the development of human civilisation contributed to the development of festivalisation (see Cudny 2014b) and how the evolution of human culture led to an increase in the variety and number of festivals all over the world.
4. To present the main factors which led to the mass development of festivals, known as the festival boom. This aim is strongly related to the previous one. The book presents briefly the main stages and factors of the development of festivals and festivalisation, especially the new ones, which appeared or developed in the last decades of the twentieth century. In this case, the phenomena and processes known as the cultural turn, cultural economy or experience societies, which greatly contributed to the development of festivalisation at the turn of the twentieth and twenty-first century are described. The factors which emerged in the twentieth century in the developed countries have recently filtered into the developing countries and contribute to the further development of festivals in the world.

5. To define the main types of festivals according to different typological criteria. The variety of festivals is extraordinary due to the many centuries of development, as well as the multifaceted character of human culture which festivals take from. Understanding how diversified the festival phenomenon is, what kind of events can be and cannot be regarded as festivals is extremely important.
6. To describe the main research methods used in the scientific analysis of festival events. Presenting the methods of studying festivals and their development, as well as their impacts on the surrounding space, is an extremely significant element of the book. It is a major objective, from the practical and theoretical point of view. On the one hand, university researchers or students may take advantage of the review of the basic research methods presented here. On the other hand, these methods are described in such detail and supplemented with so many references to literature and case studies that they can be also used by practitioners concerned with this type of events. They may use the description of the methods for their own analysis of festivals and first of all the impacts they exert on the multidimensional urban space.
7. To present the most important research themes in the analysis of festivals, including the themes found in event studies or geography. We deal here with a review of the main research approaches presented from several perspectives. Festivals are discussed from the point of view of event studies (see Getz 2012) or human geography (see Cudny 2014a). It is an important objective as the reader learns how festivals are approached in different scientific disciplines.
8. To explain the concepts of festivalisation and geographical space as well as to present the concept of urban space. Festivalisation is understood in various ways in science (see Richards and Palmer 2010; Roche 2011). Therefore, like in the analysis of festivals, it is important to specify this concept. Considering the topic of this book, it is equally important to specify the concept of space, urban space and its types as well as the relations between festivalisation and space.
9. To present the main types of impacts exerted by festivals on urban space. It is one of the most important objectives of this publication, which occupies a considerable part of one of the chapters (Chap. 4). First, the urban space is defined on the basis of the geographical definition and typology of space (see Thrift 2003). Next, the book presents the influence of festivals on different aspects of measurable and immeasurable urban space. It discusses various types of impacts, both positive and negative, quoting many practical examples.
10. To specify the concept and define the types of festival spaces. In this case, the aim is to identify and describe different kinds of spaces which the festival creates and is present in, that is to define festivalisation spaces, specify their types and quote some practical examples.
11. To describe the phenomenon of festival tourism, which is one of the most important elements shaping urban space in the context of festival development (see Cudny 2013). Here, the aim is to specify the concept of festival tourism

and establish why this type of tourism has such a strong influence on space, its inhabitants and the economy.

12. To present the concept of a post-socialist city. This aim is important due to the fact that this book contains a chapter regarding the influence of festivals on the space of Łódź—a Polish post-socialist and post-industrial city. The main transformations taking place in this type of cities in recent decades, after the fall of communism in Europe, are presented. Transformation processes in post-socialist countries are currently extremely significant, not only in Europe but globally. They largely concern urban centres, and the changes are multidimensional and similar to those which took place in western cities earlier, in the second half of the twentieth century.
13. To present the location and history of Łódź—a Polish post-socialist city. Because of the extended case study which includes the festivalisation of the urban space of Łódź, it was decided to present this city in more detail. Therefore, one of the objectives of the publication is to discuss the location and history of Łódź. Achieving this aim lets the reader understand the development of Łódź culture, Łódź festival events and the transformations they underwent after the fall of communism in Poland and the city after 1989.
14. To present the impacts of festivals on the multidimensional space of a post-socialist city, using the example of Łódź. After conducting a multifaceted analysis of festivals and festivalisation understood as the influence of festivals on urban space, their main impacts are presented, using the case study of Łódź—a city chosen for a more detailed analysis.
15. To present a model of the influence of festivals on managing modern urban space. In this case, the point was to concisely present the conclusions to the detailed analysis of the phenomenon, in the form of a model showing the various impacts of festivals on the development and management of urban space.

It was decided to adopt a research hypothesis, which is going to be verified in the discourse presented in the book. The second half of the twentieth century brought important changes in the approach to culture and its role in the space of contemporary cities (e.g. the cultural turn, creative and cultural economy). The significance of culture for social development, generating profits for cities and regions as well as creating new jobs started to be appreciated. Many cities in the developed countries in Europe and other parts of the world started to implement development and transformation strategies based on the cultural and entertainment function. Such strategies started to be introduced in many post-industrial cities as well, where the economic (industrial) base had been weakened by the process of deindustrialisation. As a result, the sector of events and festivals flourished. They were used for developing urban economy and satisfied people's need for culture and entertainment, as well as generated festival tourism. In recent decades, festivals have been performing increasingly significant social and economic functions in cities. A result of their rapid development was also the formation of a number of spaces in which festivals function and which they influence. In the 1990s, similar processes were observed in the European countries where communism had collapsed. This resulted in deindustrialisation, internationalisation of economy and globalisation in post socialist cities. In many of them, policies similar to those used in the West started to be implemented, called festival-led regeneration strategies. An example of such a city is Łódź—a Polish post-industrial and post-socialist city.

In order to write this book, a number of different research methods were used. A research method is an array of typical and repetitive ways in which reality is studied, their aim being the analysis of empirical data as a means to obtain rational answers to the researcher's questions (see Runge 2007). The research methods used in the work included the analysis of literature and Internet sources. It involved exploring all kinds of databases (e.g. Web of Science, Scopus, GoogleScholar and GoogleBooks) as well as online encyclopaedic services and other sources in order to find answers to the theoretical questions posed in the book. The idea was to identify the most frequently quoted definitions of festivals, describe their main types and present their history. Moreover, during the analysis of literature and Internet resources, many practical examples were found that illustrate the problems presented in the book. The author also made use of his own studies, such as participant observation conducted during many festivals in Poland (Łódź or Warsaw) and abroad (Great Britain, Sweden), a query at festival offices and interviews with festival organisers. The last two methods were also used to collect data and analyse festivals held in Łódź. Moreover, in this case, another method of social study was used, i.e. a questionnaire survey conducted among the participants of selected Łódź festivals and among the inhabitants of Łódź, concerning their perception of the events organised in the city. The last of the described methods were field studies which allow to collect research material that proved very useful in the wide-ranging and multifaceted analysis of Łódź festivals. The results of these studies had been published in the form of a series of articles before and are presented in Chap. 5 of this book. It was the author's earlier research and publications that formed the basis for this chapter presenting a case study of the influence of festivals on the space of a post-socialist city (Łódź). The reader will also find here a detailed description of the research methods used in festival analysis, also as regards the events referred to in Chap. 3.

The book consists of a preface, a biographical note about the author and six chapters, followed by a summary. In the preface, the reasons for writing this book are presented. The Introduction is the first factual chapter presenting the aims of the work, the hypothesis, methods and the construction of the work. In the next chapter, the concept of culture and festival definitions taken from various sciences, including anthropology, sociology and geography, is presented. This chapter presents the perception of festivals in contemporary science and the key characteristics distinguishing them from other events, such as business or sports events. Chapter 2 characterises the advent of festivals and their historical development. In antiquity, festivals were an emanation of the culture and religion of primitive tribes. Later (in Greece or Roman Empire), they developed and became more diversified as a result of the human culture development. This chapter covers the issues of the historical development of festivals as the effect of a larger amount of free time and an increase in average income, as well as the emergence of so-called experience societies in the twentieth century. This chapter also includes the description of the main factors of festival development and the features of the main types of festivals presented in the literature on the subject.

Chapter 3 also has a theoretical character as it describes the main research themes and the most important methods used in the scientific analysis of festivals. In its first part, the main research themes, e.g. the social, cultural, economic and spatial aspects, of the study of festivals are placed. The next part of this chapter contains a detailed description (with practical examples) of methods used in the study of festivals. The methods were divided into laboratory, field, quantitative and qualitative, and described as such.

Chapter 4 refers to the concept of festivalisation and the basic types of impacts exerted by festivals on urban space. First, the basic concepts such as geographical space and its types, urban space, or festivalisation and its features are described. Further on, the most important impacts of festivals on urban space divided into tangible, intangible, positive and negative are presented. The next part of this chapter concerns festival tourism. It contains the characteristics of the concept of tourism and its classifications. The issue of festival tourism as a type of tourism is described, and its influence on tourism space is presented. In the last part of Chap. 4, festivalisation spaces are described. They are urban spaces which festivals influence most strongly, e.g. festival centres as spaces of permanent festivalisation, or theatres, museums, streets and squares occupied temporarily for festival purposes, etc.

Chapter 5 presents a detailed case study of the impacts festivals exert on the multidimensional space (material and immaterial) of a post-socialist city. The example used here is Łódź—a large Polish post-industrial and post-socialist city. The analysis is based on the author's long research on the role of festivals in the development of this urban centre. The first part of this chapter contains a description of Łódź as a city—its location, history and contemporary socio-economic characteristics. In the second part, the main Łódź festivals are characterised, presenting them from the quantitative and qualitative perspective. This part of the book includes the organisers' opinions regarding the possibility of organising festivals in Łódź, as well as the opinions of Łódź inhabitants about the urban festivals, collected during the questionnaire surveys conducted earlier.

Chapter 6 is the summary of the book, which includes the main conclusions and a proposition of a model illustrating the analysis presented in the book and the role of festivals in the development of urban space.

References

Amin A (ed) (1994) Post fordism: a reader. Blackwell Publishers Ltd., Oxford

Cudny W (2013) Festival tourism—the concept, key functions and dysfunctions in the context of tourism geography studies. Geografický časopis 6(2):105–118

Cudny W (2014a) Festivals as a subject for geographical research. Geografisk Tidsskrift-Danish J Geogr 114(2):132–142

Cudny W (2014b) The phenomenon of festivals—their origins, evolution and classifications. Anthropos 109(2):640–656
Davies WKD (2015) Festive cities: multi-dimensional perspectives. In: Davies WKD (ed) Theme cities: solutions for urban problems. Springer, Chaam, pp 533–561
Florida R (2002) The rise of the creative class. Basic Books, New York
Getz D (2010) The nature and scope of festival studies research. Int J Event Manag Res 5(1):1–47
Getz D (2012) Event studies. Theory, research and policy for planned events. Routledge, London and New York
Jacobs MD, Weiss Hanrahan N (2005) The Blackwell companion to the sociology of culture. Blackwell Publishing Ltd., Malden
Męczyński M, Konecka-Szydłowska B, Gadziński J (2010) Poziom rozwoju społeczno gospodarczego i klasyfikacja małych miast w Wielkopolsce. Adam Mickiewicz University in Poznań, Poznań
Ondoš S, Korec P (2008) The rediscovered city: a case study of post-socialist Bratislava. Geografický časopis 60(2):199–213
Pacione M (2009) Urban geography: a global perspective. Routledge, London, New York
Parysek JJ (1995) Rozwój lokalny: zagospodarowanie przestrzenne i nisze atrakcyjności Gospodarczej. PWN, Warszawa
Parysek JJ (1997) Podstawy gospodarki lokalnej. Wydawnictwo Naukowe UAM, Poznań
Richards G, Palmer R (2010) Eventful cities: cultural management and urban revitalization. Butterworth-Heinemann, Oxford
Rifkin J (2000) The age of access: the new culture of hypercapitalism, where all of life is a paid-for experience. J.P. Tarcher-Putnam, New York
Roche M (2011) Festivalization, cosmopolitanism and European culture: on the sociocultural significance of mega-events. In: Giorgi L, Sassatelli M, Delanty G (eds) Festivals and the cultural public sphere. Routledge, London–New York, pp 124–141
Runge J (2007) Metody badań w geografii społeczno-ekonomicznej – elementy metodologii, wybrane narzędzia badawcze. Wydawnictwo Uniwersytetu Śląskiego, Katowice
Schulze G (1992) Die erlebnis gesellschaft. Kultursoziologie der gegenwart, Campus, Frankfurt
Słodczyk J (2001) Przestrzeń miasta i jej przeobrażenia. University of Opole, Opole
Smith A (2012) Events and urban regeneration: the strategic use of events to revitalise cities. Routledge, London
Sýkora L, Bouzarovski S (2012) Multiple transformations: conceptualising postcommunist urban transition. Urban Studies 49:43–60
Thrift N (2003) Space: the fundamental stuff of geography. In: Holloway SL, Rice S, Valentine G (eds) Key concepts in geography. Sage, London, pp 95–108
Urry J (2007) Mobilities. Polity Press, Cambridge, Malden
Wojtasiewicz L (1997) Czynniki rozwoju lokalnego-nowe ujęcie metodologiczne. Biuletyn KPZK PAN 177:7–18

Chapter 2
The Concept, Origins and Types of Festivals

Abstract This is first of conceptual chapters of the book, presenting the basic ideas connected with the festivals their genesis, development and their diversity. The chapter presents definitions of the festival taken from various sciences, including anthropology, sociology and geography. The perception of festivals in contemporary science and the key characteristics distinguishing them from other events, such as business or sports events, are presented. The chapter also includes the description of the advent of festivals. In antiquity, festivals were an emanation of the culture and religion of primitive tribes. Later, they developed and became more diversified as a result of cultural advancements, among other things. This chapter covers the issues of the historical development of festivals due to an increase in the amount of free time and average income, as well as the emergence of so-called experience societies in the twentieth century. This part of the book includes the description of the main factors of festival development. Strong focus is given to the issues shaping the popularity of festivals in the twentieth century and creating so-called festival boom. Presentation of the basic types of festivals is also one of the main aims of this chapter. This chapter also characterises their typologies based among others on attitude to religion, seasonality, form of organising and financing the event, structure of festival visitors, theme, etc.

Keywords Festivals and events · Festival definition · Festival's genesis · Development factors of festivals · History of festivals · Typology of festivals

2.1 The Concept of a Festival

Festivalisation understood as a development of festivals and their influence on people and the surrounding space is not a new phenomenon. It is closely connected with the development of human culture, which dates back to the historical roots of the human kind (Klein and Blake 2002). Due to the close relationship between festivals and culture, it was decided to start the chapter by precisely defining the latter.

W. Cudny, *Festivalisation of Urban Spaces*, Springer Geography,
DOI 10.1007/978-3-319-31997-1_2

Culture is an extremely complicated concept of multiple meanings. Its types include idealistic and materialistic, high and low, regional, local and global culture. The term derives from the Latin word meaning cultivation or education. The word culture was originally used to define the process of plant growing and animal breeding controlled by man. Analogically, it started to be used with reference to the development and self-improvement of people (Williams 1976). In ancient Rome, the term referred to the shaping of the human mind (soul). One of the first to use the word "culture" in this sense was the Roman philosopher, Cicero (cultura animi). He believed that shaping the mind through contact with various fields of philosophy and art allows a person to achieve spiritual balance and harmony with the universe (Corneanu 2011). In Renaissance, culture was commonly identified with advancements in literature, philosophy, legal order, as well as art and science, which was thought indispensable for the proper social development. Later on, primal tribes started to be distinguished from civilised societies, where these elements of culture reached a higher level of development (see Tylor 1871; Kelley 1996).

The main components of human culture included religious beliefs, which E. Durkheim (a major classical sociologist) considered to be among the most important elements. He believed that they significantly influenced other parts of human culture, such as customs, law, family or social hierarchy (see Alexander 1988). J.H. Turner claims that culture is a symbolical phenomenon which involves creating and passing on symbols used by people for communication and expression of emotions (Turner and Stets 2005).

In the nineteenth century in Germany, culture was identified with all human achievements, i.e. with the whole human civilisation (Williams 1976). It was divided into material culture, understood as real, physically present, tangible products and non-material culture, which consists of non-physical human achievements, e.g. faith, art, customs, religious and social practices. Culture was believed to have been formed earlier, determining the lives of individuals and social groups. Later, it was approached differently, as a dynamic phenomenon, created by people living at a given time in a given area. Culture was also understood as the whole human knowledge influenced by social interactions, sometimes as a way of life, a realm of values, beliefs and symbols. Many culture researchers currently believe that the phenomenon depends on the time and space in which it occurs, so it is changeable and unpredictable (Robertson 2000, pp. 37–40).

According to contemporary dictionaries, culture is "the whole material and spiritual achievement of humanity, as well as all values, rules and norms of coexistence adopted by given communities; all that is created thanks to people's work, that is the creation of their thoughts and activities. Culture is divided into material, whose range corresponds to the concept of civilisation, and spiritual, which comprises all creations and achievements in the field of art, science and morality, functioning in the form of pieces of art, beliefs, customs, as well as commonly accepted values, such as truth, justice, freedom and equality (http://portalwiedzy.onet.pl/69310,,,,kultura,haslo.html).

In the twentieth century, culture became a crucial phenomenon, not only socially, but also economically and politically. Its role was growing in the subsequent decades of the twentieth century, but it was the strongest towards the end of the century. In the 1970s and 1980s, people started to talk about the cultural turn in politics, economy and science (see Barnett 1998; Ray and Sayer 1999; Barnes 2001). Culture was treated as a product, which brings material and promotional benefits. Its political role increased as well, because the politicians' influence on the economy got weaker due to globalisation and the growing role of global corporations. Thus, they tried to intensify the discussion concerning culture and its influence on values, the society and economy. These elements often continued to depend on politicians at different levels, e.g. due to the public financing of culture. In this way, they were easier to manipulate than economic processes, and the politicians could boast of successfully creating culture-related events. On the other hand, more and more significance was being attached to the roles of culture in social life, as the twentieth century and twenty-first century have been a period of intensive globalisation, international migrations and clashing of the cultural patterns of various societies. Cultural phenomena, such as multicultural societies or movements related to the women's emancipation and equality of sexual minorities, are becoming more and more significant, politically and socially.

The role of culture in the contemporary world increased also due to the change from the industrial to the post-industrial and post-Fordist economy. The twentieth century brought the deindustrialisation of the developed countries and societies whose economies were based on services, including those related to culture and entertainment. The importance of cultural phenomena as products was appreciated, as well as their role in forming the creative class (see Florida 2002a, b). All this was noticed by many entrepreneurs, managers and politicians on the national and local level, and as a result, culture became a valuable economic component, used for the socio-economic stimulation of regions also those in crisis. It was included in the marketing strategies of countries, regions and cities and used for their economic development, promotion and creation of a positive image. There appeared numerous culture-based products, which play an increasing role in the local, national and global economy. The growing economic importance of culture started to be referred to as cultural economy (see Ray and Sayer 1999; Gay and Pryke 2002), which played a particular role in urban regions, where it was defined as the cultural economy of cities (Scott 2010).

The cultural processes described above include the development of festivals. They have always been a major component of human culture, being connected with religious beliefs, from primitive ones (e.g. animism, shamanism, totemism) to the religions currently dominating the world (e.g. Christian religious festivals). Festivals are a part of the non-material culture, as they present art, customs and cultural symbolism. They may be an emanation of the local or regional culture (small, e.g. community-based or regional festivals), but also of the global culture (large-scale film or music festivals). They represent high culture (e.g. classical music, ballet festivals), but are also organised by the creators of pop-culture. As pointed out by Cudny (2014a), they are used in politics, too, e.g. for the promotion of a political party or individual politicians, perform important social functions and play a growing economic role. Festivals are an important element of most aspects of culture. Therefore, the development and the growing importance of culture are followed by the increasing role of festivals in the contemporary world.

The cultural turn described earlier had an effect on the academic milieu as well. Researchers representing many sciences, especially humanities, started to focus more on the multidimensional human culture. Some of the disciplines which so far had been exploring culture redirected their approach to it and took into consideration its multidimensionality, pluralism, changeability, as well as the role of individuals and social groups in its formation (Jacobs and Weiss Hanrahan 2005; Hawkins 2013). Researchers' increased interest in culture was also reflected in the development of event and festival studies which could be observed in various sciences in the last decades of the twentieth century.

In order to describe the concept of festivals and the related phenomena, we must start from explaining the term "events", because festivals are very frequently treated as such. There are many different types of events, classified according to their theme, scale, the way they are organised, etc. The very word "event" is of Latin origin, and there are some similar expressions connected with it:

1. *eventus* meaning a result, effect, success;
2. *eventum* meaning a happening, manifestation, outcome;
3. *evenire* meaning to appear, to happen; and
4. *venire* meaning to come, to sell, to be sold.

There are also additional, culture-related meanings of this term, which include a gathering or social activity, something that happens at a given place and time, competition, adventure and occasion (Tara Lunga 2012, pp. 761–762).

Among the first comprehensive definitions of events, there are those provided by Hall (1989), Goldblatt (1990a, b) and Getz (1991). Hall (1989) states that hallmark tourist events are major fairs, expositions and cultural and sporting events of international status which are held on either a regular or a one-off basis. The primary function of a hallmark event is to provide the host community with an opportunity to secure high prominence in the tourism market place. According to Golblatt (1990a, b cited in Tara Lunga 2012, p. 763), an event is "a unique moment in time celebrated with ceremony and ritual to satisfy specific needs". Goldblatt (2000, p. 3) wrote that "Robert F. Jani, the first director of public relations at Disneyland described the Main Street Electric Parade as 'a special event' in 1954… He further explained, 'I suppose it (special event) is that which is different from a normal day of living." In one of his later works, Getz (2008, p. 404) claims that "Planned events are spatial—temporal phenomena, and each is unique because of interactions among the setting, people, and management systems—including design elements and the program. Much of the appeal of events is that they are never the same, and you have to be there to enjoy the unique experience fully; if you miss it, it's a lost opportunity". It can be seen from the review of concepts above that the basis of every event is a properly designed, interesting, unusual and unique experience. Without it, the event (including festivals) could not function because it would not attract the audience (Getz 2012, p. 8).

With reference to the work by Arcodia and Robb (2000), Cudny (2014b, p. 641) divided events into several types, according to their theme and scale: events (mega events, major events, hallmark events, signature events, special events), festivals (community celebrations, community entertainment, historical commemorations, multicultural celebrations, seasonal events, religious celebrations) and a variety of events connected with Meetings, Incentives, Conventions and Exhibitions (the MICE sector). Jago and Shaw (1998) divided events according to their range and size. They distinguished among minor events, major events, hallmark events and mega events and festivals. Getz (2005) created a similar typology, dividing events according to their size and importance for the society and economy into the following types:

- local: they are in low demand, of little economic and social importance, satisfying mainly the needs of small, local communities,
- regional: they are in medium demand and of medium importance for tourism development,
- periodic hallmark events: they are repeated over time, in great demand, of considerable importance for the socio-economic growth and tourism development, and
- occasional mega events: one-off events, which are in great demand and of huge importance for the socio-economic growth and tourism development.

Another typology, based on the theme of events, can be found in Getz' next work (2008, p. 404), where he divides special events into the following types:

- Cultural celebrations: festivals, carnivals, commemorations and religious events;
- Political and state: summits, royal occasions, political events and VIP visits;
- Arts and entertainment: concerts and award ceremonies;
- Business and trade: meetings, conventions, consumer and trade shows, fairs and markets;
- Educational and scientific: conferences, seminars and clinics;
- Sport competitions: amateur/professional and spectator/participant;
- Recreational: sport or games for fun; and
- Private events: weddings, parties and socials.

Festivals not only are the domain of sociological or anthropological research, but undergo a thorough analysis in event studies as well (Getz 2010), where they are treated as a separate type of events, also by Arcodia and Robb (2000). Getz (2008), too, approaches them as a separate type of events, related to culture. In the classification based on the scale and importance presented by Getz (2005), festivals can be practically included in every type of events. There are small, local festivals (neighbourhood festivals), regional festivals and large-scale events (hallmark or mega events), attracting audience from the whole country or even abroad (e.g. the Cannes Film Festival and Woodstock).

Having presented the definitions and typologies of events, as well as the role played by festivals, it is necessary now to give a precise definition of the festival. It should be stressed that it is not a simple concept, easy to present. For instance, according to the online Oxford English Dictionary (http://www.oed.com/view/Entry/69567?redirectedFrom=festival+#eid), the word "festival" derives from "old French *festival*, *-vel*, medieval Latin *festivalis*, Latin *festivus*" and is an adjective referring to:

- a feast, befitting a feast-day; glad, joyful, merry;
- the time of festive celebration, a festal day. Also occasionally, a festive celebration, merry-making. Also, to hold festival, to keep festival, to make festival, to proclaim festival;
- a musical performance or a series of performances at recurring periods, mostly of 3 years, e.g. the Handel Festival, the Birmingham and Norwich Festivals (see Grove Dict. Music at Festivals). Also applied to a series of films, theatrical performances, etc.

As Cudny (2014b, p. 642) wrote, there are several dictionary definitions describing festivals. He cited, among others, the definition from Encyclopaedia Britannica, where a festival (also feast) is "a day or period of time set aside to commemorate, ritually celebrate or re-enact, or anticipate events or seasons—agricultural, religious, or socio-cultural—that give meaning and cohesiveness to an individual and to the religious, political, or socio-economic community (http://www.britannica.com/EBchecked/topic/203113/feast)".

In the contest of event studies, Getz (2005, p. 21) defined the festival as a "themed public celebration". In this case, we deal with a definition describing the festival as any event which is public and concerns any issue interesting to a given group of people. The authors of works in the field of event studies often quote a definition presented by Falassi (1987, p. 2), according to which "Festival commonly means a periodically recurrent social occasion in which, through a multiplicity of forms and a series of coordinated events, participate directly or indirectly and to various degrees all members of the whole community, united by ethnic, linguistic religious, historical bonds, and sharing a worldview". A. Falassi is an Italian sociologist who worked in the USA for many years, investigating the local culture and folklore. His research got him interested in festivals and their role in social life (see Falassi 1980). In the definition quoted above, a considerable significance is attached to the social role of the festival as an organised phenomenon, in which all members of a given community take part. In this case, we deal with the festival as a phenomenon integrating the community or even a nation, based on its common values (national, language, religious and historical).

The definition puts the strongest emphasis on understanding the festival as a form of expressing the community spirit on the social and national level, due to taking an approach typical of sociology or culture anthropology. In this sense, this definition has a lot in common with the classical understanding of the feast or festival in anthropology and sociology, which was presented by E. Durkheim and

J. Frazer at the turn of the nineteenth century and twentieth century, though a similar understanding of this phenomenon can be also found in more recent anthropological (Cruikshank 1997; Roemer 2007, 2010) and sociological works (Duvignaud 1976). In the early works of this type, those phenomena were linked to the cultures of primal tribes (e.g. African or Australian), where such events were connected with the beliefs, culture or tradition of given tribes, and later also certain social groups.

Festivals of this type are still an important but not the only segment of the festival market. Therefore, from the modern point of view, this definition seems to be incomplete. For instance, nowadays, many festivals are not connected with a specific place or community; they are held at different locations and refer to global culture (e.g. some film or music festivals). In this sense, they do not refer to specific nationalities or social groups sharing the same values, language or history. Moreover, a particularly important element is the experiences factor, generating demand for festivals. The definition quoted above does not refer to this issue. Many people searching for extraordinary experiences which might enrich their lives more and more often attend some festivals, even if they do not feel emotionally attached to them for historical or cultural reasons. These are new, unknown experiences for them, and thus, they are interesting and evoke a special kind of excitement. What is more, the definition presented above lacks references to the influence of events on economy and space, which is currently a very significant issue.

A slightly different definition of the festival was suggested by a Belgian anthropologist working in France, Piette (1992, cited in Cudny 2014b, p. 642), who believes that "…festival is portrayed as reinforcing established society. The antithetic behavior of the festival is said to destroy social convention in order to reinforce it. Thus, the festival is displaced from its proper logic, that of ritual, rules and regulations, play and ambivalence". This definition emphasises the aspect of festivals connected with disturbing the social order established in a given group. It is an important function of events which the same author calls transgression festivals. They enable people to get away from the everyday routine, leave the everyday ordinary social roles and move to the world of hedonistic, often uncontrollable fun. It is not important what social status people have in everyday life. Meeting at such a festival, they make new interactions, ignoring (temporarily) their social roles (see also: Caillois 2001). This important function of festivals was observed already in primal tribes or ancient cultures, e.g. in Greece (the Dionysia) or Rome (Saturnalia), as well as during the medieval carnival. It is still present in festivals organised for the youth or connected with different kinds of subcultures (e.g. Woodstock) (Cudny 2014b). They are events which accumulate and represent social rebellion and protest. They may be qualified as a particular kind of events organised by the people against the establishment and the social order it represents (see Falassi 1987).

In geographical works, festivals are defined in a variety of ways. As Davies (2015, p. 535) wrote "Festivals are distinctive because they take people outside their normal behaviours in time and space. They provide unusual activities and evoke feelings and emotions that are very different to the regular and material routines of the workday. Traditionally most took place in spaces that are either normally used for other activities such as roads or were empty spaces, but later become the exclusive sites for the period of the event—sites usually temporarily transformed by decorations and events that add to the sense of occasion and the separateness of the experiences gained in that space". An American geographer studying festivals defined them as "formal periods or programs of pleasurable activities, entertainment, or events having a festive character and publicly celebrating some concept, happening or fact" (Janiskee 1980, cited in Quinn 2009, p. 485). Waterman (1998, p. 55) wrote that "(…) festivals transform landscape and place from being everyday settings into temporary environments—albeit with permanent identities—created by and for specific groups of people, (…) art festivals contribute to both the production and consumption of culture (…), provide a means whereby groups may attempt to maintain themselves culturally, while presenting opportunities to others to join that group. Festival is also an occasion for outsiders (sponsors, subsidizers) to endeavour to force or to lead the group towards an acceptable course for the continuity of its culture. This is no esoteric aesthetic topic; arts festivals have become events of sociological and geographical concern".

In the work by Quinn (2005, p. 3), the definition formulated by Isar (1976) is cited, where festival is "something exceptional, something out of the ordinary (…), something that must create a special atmosphere which stems not only from the quality of the art and the production, but from the countryside, the ambience of a city and the traditions (…) of a region".

Australian geographers define festivals in the following way:

1. "Festivals are enjoyable, special and exceptional, sometimes the only time of celebration in small towns. Festivals are full of rituals of entertainment, spectacle and remembrance, and they bring people together. Most people participate for enjoyment, something different and the pleasure of coming together" Gibson et al. (2011, p. 3).
2. "To qualify for inclusion as a festival, an event had to meet at least one (and preferably more than one) of the following criteria: use of the word 'festival' in the event name; it being an irregular, one-off, annual or biannual event; emphasis on celebrating, promoting or exploring some aspect of local culture, or being an unusual point of convergence for people with a given cultural activity, or of a specific sub-cultural identification" (Gibson and Stewart 2009, p. 6).

According to the Polish human geographer Cudny (2014b, p. 643), "a festival is an organised socio-spatial phenomenon, taking place at a specially designated time, outside the everyday routine, shaping the social capital and celebrating selected elements of human tangible and intangible culture."

Geographical definitions emphasise the unique character of festivals, their role in celebrating culture and the fact that they are meeting places for people with specific cultural interests. Besides, some of them (e.g. those quoted by Waterman 1998 or Cudny 2014b) stress the role of festivals as a spatial phenomenon, which temporarily transforms the space it occupies. Such a transformation is often temporary, which means that the space is occupied for the festival purposes for a short period (a few to several days), over which it is completely differently managed. An example of that could be the space of a street in the city centre—normally a free public space used for walking and shopping. During a festival, it is taken up by a stage, stalls, spectators and festival staff (Fig. 2.1). It seems that the spatial aspect of festivalisation is not sufficiently accentuated in geographical definitions.

The influence of festivals on the space and the changes they cause in different dimensions of geographical space should undergo detailed analysis in geography. It should also be more strongly reflected in the geographical definitions of festivals, where we should refer to the concept and division of geographical space. In the context of this publication, the aspects of urban space and the influence of festivals on this particular type of geographical space are so important; thus, they are not discussed in this part of the book but presented in detail in another chapter (see Chap. 4).

Cudny (2014b) distinguished several social and cultural components of a typical festival. They can be divided into two main groups (Table 2.1):

1. those connected with the organisation of the festival event itself and
2. those connected with the impacts of the event on culture and the community.

As for the organisation of a festival event, it takes place at a specially designated place and time, is a public celebration and has a leading theme. A festival is an event based on extraordinary experiences (e.g. cultural) which during an event occupies specified place for a specified period of time. Moreover, a festival can be either a one-off or a regular event and may include a competition. As for the other

Fig. 2.1 The main street of Łódź (Piotrkowska Street) during a street festival in 2014. *Source* Author's photograph

Table 2.1 The main components of a festival

Elements connected with the organisation of a festival	Elements connected with the impacts of the festival on culture and the community
It takes place at a specially designated place and time (occupies this space for a specified period of time), outside everyday duties	It develops social identity and social capital, e.g. through celebrating values important for a given community, and the skills acquired during the festival
It is often combined with a competition for the best creation (e.g. a film) or an overview of an artist's works	It develops interpersonal relations through contacts made and reinforced while it lasts. It concerns groups of friends, family, colleagues and school friends
It is a public event, containing an element of fun	It is a social event based on extraordinary experiences; it is generated by social needs; the participants are community (local, regional) members; it has numerous social impacts
It is an organised event with a leading theme	It is a culture-related event. It is rooted in culture, shapes and presents it; it is a place and time of celebrating and consuming culture
It may be a one-off or regular event	It is a part of broadly understood human cultural heritage, particularly its intangible part

Source Cudny (2014b, p. 644), modified

group of festival components, they show that it is a part of human culture (cultural heritage). It is undoubtedly a social event making interpersonal relations more profound, integrating the community as well as building social identity and social capital. The latter consists of interpersonal ties, relationships, acquaintances and cooperation within a community (Table 2.1).

When analysing the concept of a festival, we should also consider the environment in which modern events of this type are functioning, as they take place in specific, socio-economic and environmental settings. The whole existence of a festival, i.e. its organisational structure, budget, individual events and impacts on the surroundings, is determined by the elements of these surroundings, such as the organisers, the participating artists, sponsors, politicians (supporting or blocking the event) and the audience. Moreover, the festival is also influenced by its economic environment, infrastructure (used for the purposes of the event), the media, local or regional culture, the geographical surroundings of the festivals, etc. (Fig. 2.2) (Hauptfleisch 2007).

An interesting approach is presented by Gibson and Stewart (2009), who refer to the festival as "a point of convergence". In this case, we deal with a feature of the festival discussed by geographers (Quinn 2005; Cudny and Ogórek 2014) or culture experts (Valck 2007), who present festivals as places of meetings, contacts among people interested in a certain form of culture and art, as well as people working in the fields related to this kind of activity (film-makers, musicians, sponsors, agents,

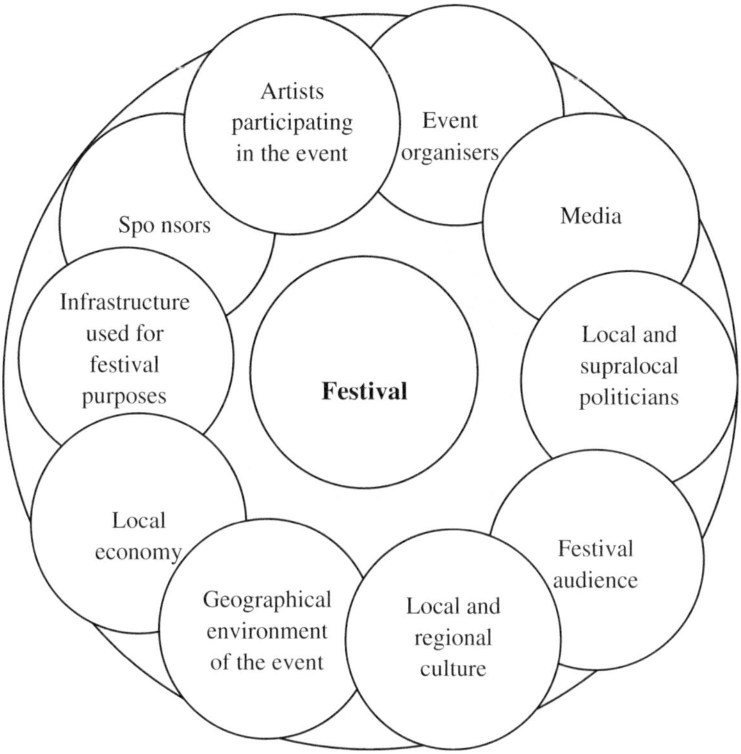

Fig. 2.2 Elements of the festival environment. *Source* Own elaboration on the basis of Hauptfleisch (2007, p. 43)

journalists, etc.). It is important to present the festival in the contexts of the social network theory and the obligatory points of passage (or convergence), which are elements of such networks. They are particular points of social and business contacts and a part of more extensive relational links within the social network of a given community. The social network theory has a long tradition. It is based on the assumption that there are various links among different actors in a society, which form functional and spatial systems within the social space. Such networks include relationships between the persons who create them and institutions. These relationships are very important because social networks enable their participants to maintain personal and virtual contacts, to exchange information and ideas, or to cooperate creatively on specific projects (Wasserman and Faust 1994).

Culture and its actors—the organisers of cultural—artistic events (including festivals), the artists and institutions cooperating with them (galleries, theatres, cinemas, local authorities, foundations, etc.) and the audience and the volunteers form specific relational networks among them, too. In culture-related networks, festivals are a particular type of meeting points (see Quinn 2005; Valck 2007). According to Valck (2007), film festivals are very important for the world cinema.

Without them, the film network could not exist and would fall apart. That is why Valck called film festivals obligatory points of passage for the film industry.

Many art festivals are such obligatory points of passage for artists, businessmen, journalists, spectators and volunteers involved in a given type of art (film making, theatre, music). In this way, art festivals allow contacts between artists and producers, agents and sponsors, as well as popularise the knowledge about a given art. Finally, they are places of presenting and promoting pieces of art and discovering young talents; they are also an opportunity for artists to compete against one another, e.g. during festival competitions (for details see Cudny and Ogórek 2014).

2.2 The Origins and Development of Festivals

When discussing the development of festivals over time, we should stress that already primal peoples had rituals similar to today's festivals (e.g. religious ones) and related to the most important moments in the life of a given tribe and its individual members. The rituals had the form of games accompanying, e.g. religious holidays, births, the coming of a new year or electing a new chief or king. Festivals treated as rituals connected with such important moments in the lives of social groups were organised on different continents, e.g. in South America, Asia, Africa or Europe. One of the first festivals was also those events connected with seasonal occasions such as sowing and harvesting (Cudny 2014b). Those kinds of festive events were also seen as the first festivals by Davies (2015, p. 541), who called them temporal or seasonal festivals. More advanced forms of festivals, containing elements of higher culture or art, took place in ancient Greece. The Greek word for festival was heorte, which meant urban celebrations connected with Greek gods. Heorte involved making a sacrifice, with the participation of priests, a procession and a gathering of people who were watching the whole event and could actively take part in it as well. In this way, people were celebrating the anniversaries of the gods' or heroes' births, as well as a variety of other special occasions. Such events often included feasts, singing songs, reciting poems or theatre performances, as well as competitions, e.g. for the best piece of literary art. Thus, festivals had a strong cultural element, apart from the religious one.

Examples of such festivals are the events accompanying the cult of goddess Demeter and the Dionysia (Wilson 2006). In as early as about 1450 BC, the first sanctuary in Greece was built. It was erected in Eleusis and devoted to Demeter. The cult of Demeter was connected with rituals called the Eleusinian Mysteries. Many historians claim that an important part of those rituals was a performance, during which episodes from the myth about Demeter and her daughter Persephone (Cora) were presented. Parts of this myth, acted out by priests and followers, showed Demeter suffering after she lost her daughter and rejoicing in her return. Next, the Eleusinian Dionysia included the cult of the Greek god Dionysos. Rituals connected with this god later became a widely celebrated holiday in ancient Greece (Grzesiak 2010).

Dionysia were treated in ancient Greece as a separate holiday as well, dedicated to Dionysos (in ancient Rome known as Bacchus). Dionysos was the son of Zeus and a mortal woman; he was a god of fun, vine and wine. In the beginning, his cult had an orgiastic character, which with time was toned down and civilised. People drank wine in his honour, organised feasts and parades and wore masks presenting the gods of earth and fertility. The parades turned into stage performances of three types: comedy, tragedy and drama. Four holidays were celebrated as parts of the Dionysian cycle: the Small (Rural) Dionysia, celebrated by the population of Attica (December–January), the Lenaia (January–February), the Anthesteria (February–March) and the Great Dionysia (City Dionysia) (March–April). In about 534 BC., the Great Dionysia included the tragedy, and earlier, on the occasion of the Lenaia, the first comedy was staged in Athens (http://encyklopedia.pwn.pl/haslo/Dionizje;3892862.html).

The main Dionysia were held in the most important Greek city state of that time, i.e. in Athens. Every year, about 15 new texts were written, representing comedy, satire and tragedy. Over 100 years, about 1500 texts were created in Athens alone. Moreover, Dionysia took place in other cities of ancient Greece as well, where they were also accompanied by theatre performances and competitions (Ashby 1999). All that contributed to the development of Greek theatre, which in turn became the basis for the development of theatre in Rome at the time when Greece was under the rule of the Roman Empire.

In ancient Rome, one of the most important religious festivals was the Saturnalia, organised in honour of the agricultural deity Saturn, on 17–24 December (Cudny 2014b). The festivities were divided into the formal and the informal part. During the first one, offerings were made to Saturn. Near the Temple of Saturn, next to the Roman Forum, a ritual feast was held. During the holiday, Romans did not wear the official toga but a casual outfit. Slaves swapped roles with their masters, thus referring to the mythical golden age, when Saturn ruled and all people were equal. The holiday included feasts, during which huge amounts of wine were consumed, gambling flourished and masquerades and parades in disguises were organised. Romans used to make embellishments from evergreen plants and decorate their houses with garlands and lamps. The holiday time was free of work and school and spent visiting friends and family (Radziejewski 2006).

Ancient festivals referring to religious holidays are reflected in the modern world. Firstly, there are large religious festivals nowadays, which are organised at many places all over the world. Examples of such festivals are also described in scientific literature. They include events such as traditional festivals in Japan (Roemer 2007, 2010), religious festivals connected with ethnic minorities, e.g. in New Zealand (Tondo 2010), puppet theatre festivals connected with traditional beliefs, organised in Indonesia (Osnes 2010), or Christian festivals in Poland (Cudny et al. 2011).

We also find examples of religious festivals in Asian countries such as China and Taiwan where those festivals appeared already in Early Middle Ages. An example is The Ghost Festival, dedicated to the spirits of dead ancestors and draws on Taoism and Buddhism. This festival is still organised today in many Chinese and Taiwanese cities and rural destinations. The festival takes place on the fifteenth day of the seventh month of the year. It is a period in which the ghosts come out from the underworld and visit the living (see Teiser 1988).

Another example of a religious festival based on Buddhist traditions is a festival in Wutaishan (China). This event has been revived in recent years, among others, in order to increase tourist traffic in this region of the country. In the last two decades, we saw a revival of many traditional rituals, including a variety of cultural and entertainment events, in China. These events and rites, including religious festivals, were forbidden by the Chinese communist authorities for a long time. Currently, those events are treated as a generator of domestic and international tourism (event and heritage tourism) and a generator of revenue from the tourism sector (Ryan and Gu 2010; Szczepański 2012). We also can give the example of quite similar religious festivals held in India. These are among others Deepavali, Divali or Maha Kumbh Mela festivals. The last one is often described as world largest public gathering (bathing for purification from sin) and is attended by ca. 80–100 million people. These events have religious meanings and they attract millions of festival tourists in the country as well (Davies 2015).

On the other hand, ancient Greek festivals which were described before frequently referred to the countryside, harvest, wine and food. All these spheres are represented by modern festivals as well. Rural festivals should be understood as all the festivals which take place in rural areas; however, only some of them have an agricultural character. They date back to the nineteenth century, when farmers started to present new methods of land cultivation and new produce in the form of festivals. There are many scientific works presenting small, local festivals based on rural culture and connected with the celebration of harvests, crops, seasons of the year, etc. Such festivals were described by Janiskee (1980, 1991) in the USA, and in Australia, many examples were presented in a book edited by Gibson and Connell (2011). Later, agricultural festivals also became an occasion to present farming equipment and plant protection products, as well as a place of entertainment. They became a cultural phenomenon and a source of local community integration.

In the twentieth century, art festivals started to be organised in rural areas. They were still rural events but had lost their agricultural character, because they concerned, e.g., types of music completely unconnected with rurality (see Brennan-Horley et al. 2007). Thus, rural festivals went a long way from typically agricultural events, which may be compared to many ancient festivals related to crops and harvest and to art festivals, which often perform functions similar to their urban counterparts, though on a different scale.

Another type of festivals which may be derived directly from ancient Dionysia or Roman Bacchanalia is modern wine and food festivals. Despite the fact that they do not involve such licentious and orgiastic feasts and drunkenness as in the past, they are to a certain extent a continuation of ancient traditions. During the Dionysia, feasts were held and huge amounts of wine, the patron of which was Dionysus, were consumed. Romans adapted this cult, and Dionysus was called Bacchus. The festivities organised in ancient Rome in his honour were called Bacchanalia. Drunkenness lasting for many days was accompanied with feasts and indecent orgies. Today's wine festivals have lost this orgiastic character. They are usually held in vineyard areas, rich in wine making traditions. They are events enabling wine producers to promote and taste different wines and generating considerable tourism at many destinations all over the world. They are connected with various stages of wine growth: bud break, flowering, fruit set, veraison and harvest. They are organised in Europe, North America, Australia and recently also in Asia, e.g. China. Dozens of wine festivals are held in Australia, France, the USA, Hungary, Greece or even Poland. Thematically, similar events, also connected with alcohol consumption, are the beer festivals, with the most famous October Fest held in Munich (Hall and Sharples 2008).

On the other hand, contemporary rural destinations as well as cities more and more often organise food festivals, which attract gourmets from all over the world as a part of food tourism (Hall and Sharples 2008). There are many examples of food as well as food and wine festivals all over the world, such as the San Francisco Street Food Festival, Truffle Festival in Italian Alba, Melbourne Food and Wine Festival, Maine Lobster Festival in Rockland, USA, Pizzafest in Naples, Galway Oyster Festival in Ireland and Great British Cheese Festival in Cardiff. An interesting example is also the Tatebayashi Noodle Grand Prix Festival, an annual food festival hosted by the city of Tatebayashi in Gunma prefecture, Japan. This is the example of the event, where unique regional food triggered the development of festival tourism (see Kim 2015). It is possible to find references to ancient feasts held during Dionysia or Bacchanalia in this type of contemporary events.

As shown by Quinn (2009), festivals developed also in subsequent historical periods following antiquity, i.e. in the Middle Ages and Renaissance. In the Middle Ages, the tradition of carnival was born—the festive time preceding the period of Christian Lent. New events which appeared at that time, e.g. the medieval knights' tournaments, celebrations during visits of medieval rulers to the cities subject to them or court games, can be seen as the archetypes of many contemporary festivals. Apart from feasts, those events included cultural events, such as theatre performances, recitations and concerts. A popular form was performances given by travelling theatrical groups, which increased people's interest in culture and attracted the inhabitants of cities and villages (see Cudny 2014b).

The most significant medieval event of the festival type was certainly the carnival, whose tradition still continues (e.g. the Carnival in Venice, Cologne, or Rio de Janeiro). According to Encyclopaedia Britannica, a carnival is "the merrymaking and festivity that takes place in many Roman Catholic countries in the last days and hours before the Lenten season. The derivation of the word is uncertain, though it possibly can be traced to the Medieval Latin carnem levare or carnelevarium, which means to take away or remove meat. This coincides with the fact that Carnival is the final festivity before the commencement of the austere 40 days of Lent, during which Roman Catholics in earlier times fasted, abstained from eating meat, and followed other ascetic practices. The historical origin of Carnival is also obscure. It possibly has its roots in a primitive festival honouring the beginning of the new year and the rebirth of nature, though it is also possible that the beginnings of Carnival in Italy may be linked to the pagan Saturnalian festival of ancient Rome" (http://www.britannica.com/EBchecked/topic/96363/Carnival). In the Middle Ages, the carnival was a period of unrestricted freedom and fun, a time when social roles were swapped and conventions broken. Thanks to it, people balanced the period of formal oppression with severe everyday rules and regulations. Moreover, it was a period which enabled people to let off some steam before the approaching restrictive fasting. The carnival was a specific, short-term social revolution, a safe valve in the formalised societies of those times. It is also a typical example of the transgression festival (Eco 1984).

One of the Venice Carnival attributes was the mask, which allowed the person who wore it to hide their identity. People could indulge in all sorts of fun with impunity and take advantage of brothels and game parlours, which was officially condemned and severely punished by the authorities. Masks concealed the sex; women could dress up as men and the other way round. During the carnival, the social roles were reversed; the rich and the poor had fun together, hidden behind their masks and free from the conventions and formal laws. In numerous comedy performances, it was possible to make fun or even humiliate high-ranking church officials and rulers. It was allowed by the festival convention, which started with a publicly performed symbolical scene presenting a falling angel. The role of the king of the carnival was given to a poor man or a fool. The carnival ended with a ritual of dethroning the carnival king; the social order as well as the God's and human laws were restored. The carnival tradition has remained alive in many European countries, e.g. in Italy (Venice), Germany (Cologne) or France (Nice). As a cultural phenomenon, the carnival spread to other continents as well, e.g. Africa (Zimbabwe—Harare Carnival), South America (Brazil—Rio de Janeiro) and North America (The Caribbean, the USA—Louisiana).

The Renaissance brought development of festivals in Europe (Quinn 2009), and one of the major festivalisation centres was Venice. Apart from the famous Venetian Carnival, many other religious and secular festivals were organised there (see Muir 1981). As pointed out by Cudny (2014b), the consecutive centuries brought a development of culture- and art-related events. According to this author (citing Strong 1984), they included some types of court performances: dance, pantomime, opera singing in the seventeenth century in France or the first music festivals, such as the English Three Choirs Festival in Hereford or Handel Festival in Westminster Abbey (seventeenth century). Renaissance revived the idea of antiquity and brought back to life traditional festivals originating from the period of ancient Greece or Rome. Besides, Renaissance also developed the habit of feasting as exquisite social meeting. Feast combined with cultural and entertainment events became an important social event. It became the setting for important events such as marriages, celebrations of war victories or even coronations of kings. During feasts, the rank of its participants was shown; thus, the feast was an emanation of power and authority. At the same time, it was the place of culture celebration and the place of a social discussion and consumption (Strong 1984, 2002; for festivals in Renaissance see also: Jacquot 1956–1975).

In the sixteenth century and seventeenth century, the Protestant Reformation arose in Europe, together with the rise of Anglican Church. Important part of reformation in England was the Puritan movement, which was first the reformation of Anglican Church, a later developed into almost separate church. They had a postulate of purification of the church; thus, they promoted the austere lifestyle, based mostly on work, prayer and study of the Bible. Thus, they considered alcohol and any kind of entertainment such as gambling or theatre shows as immoral. This together with the rise of religious wars in Europe and counter-reformation movement in Catholic church led to the decline of many festivals. Later on with development of industry (in eighteenth–nineteenth century), people focused on work. This was connected with the limited amount of free time in the early industrial society and withdrawal from traditional values (e.g. religious). Those factors led to the decline of many old, traditional festive events in early industrial Europe. However, the turn of eighteenth and nineteenth centuries were also the time of rise of new national states (such as Italy and Greece). What led to the redevelopment of festivals, very often organised by the state or by the political leaders, to support various political ideas (see Davies 2015). On the other hand in the eighteenth century, the idea of the Grand Tour started to flourish among young British aristocrats. "The Grand Tour became a rite of passage for aristocratic young men. The journey typically involved three or four years of travel around Europe and included an extensive sojourn in Italy, as Rome was considered the ultimate destination for what might now be characterised as cultural tourism" (http://www.britannica.com/EBchecked/topic/1567446/Grand-Tour). Later, as the bourgeoisie developed, the Grand Tour became popular also among that social class and outside England.

Festivals were developing, and their influence on people and economy was increasing due to the advancements of the industrial age, which began at the time of the first industrial revolution. The steam engine invented in 1769 was used as an efficient device in industry as well as in transport. It was installed on steam ships (by William Symington 1801–1802 and later Robert Fulton 1807), as well as on steam locomotives (George Stephenson 1829). Thanks to faster, more comfortable transport, tourism started to develop. The appearance of the steam ship and steam engine (locomotive) as well as the development of railways, first in Europe and then all over the world, contributed to the growth of mass tourism (Gierczak 2011).

Industrialisation, which started in the eighteenth century, finally resulted not only in transport conveniences and an increase in people's average income. The next industrial revolutions brought other innovations, such as electrical power or growing mechanisation of production. Due to the inventions, the production capabilities were growing, products were becoming cheaper, and people could earn more. A class of rich entrepreneurs developed and the middle class started to form as well. Due to the civilisational advancement and man's changing attitude to life and work, a number of social benefits were introduced, e.g. paid holidays, regulations concerning working hours or retirement schemes implemented in industrialised countries. All that gave people more and more financial resources and free time.

Additionally, the nineteenth century brought other important inventions, such as the combustion engine (Gottlieb Daimler 1889), the first transatlantic lines, the building of the Suez and Corinth Canals and the development of the first large tour operator, organised by J. Cook.

In the twentieth century, planes appeared, which, after the maiden flight by the Wright brothers in 1903, became one of the major means of transport for people and goods (Gierczak 2011). All those achievements led to the development of global tourism, also its part involving broadly understood culture, festivals included. The events organised for the first time in the nineteenth century were, e.g. German Oktoberfest (beer festival), Bayreuth Festival dedicated to R.Wagner and Venice Biennale (see Cudny 2014b). Quinn (2009, p. 6) claims that "in 1859, the Handel Centenary Festival held at London's Crystal Palace was marketed as a tourist attraction with the organisers distributing 50 000 prospectuses in the European offices of the railway companies serving the Crystal Palace". In the nineteenth century, world exhibitions started to be organised, where scientific and technological achievements from all over the world were presented. They also included cultural and entertainment events, similar to today's cultural festivals (Cudny 2014b).

It was also a time when spas started to develop, performing an increasingly important cultural and entertainment function (concerts and performances for the patients). A characteristic feature of spa treatment is that after therapies the patients have a lot of free time, which they spend on entertainment and social life. In the nineteenth century, European spas became centres of social life, or even gambling. Members of the European elite visited them not only for health reasons, but also for entertainment, to make new friends and experience culture and art. They were the audience of numerous concerts, theatre performances and regular festivals. Theatres, concert halls and amphitheatres in many spas became event venues, and stages were built, e.g. in spa parks.

Currently, spas are an important festivalisation space. A good example of that is Krynica Zdrój in Poland where Jan Kiepura Festival is held. Kiepura was a famous opera singer and actor working in Poland, Austria, Italy and the USA. One of his favourite places in Poland was Krynica Spa, where he built a private sanatorium, called Patria. He often visited that destination, singing at events organised there. To commemorate that artist, in 1967 the first Jan Kiepura Festival was held in Krynica Zdrój. The event combines several arts belonging to high culture, including ballet dancing shows, opera singing and theatre performances. About 1000 artists from Poland and abroad come to the festival every year, and it is attended by over 60 000 visitors (Jędryka 2012).

Another example of a festival organised in a popular spa is The Royal Theatre Festival in Spa. Spa is a well-known medicinal destination in Belgium, situated in the Ardennes. The curative properties of the mineral waters in that area were famous already in the times of the Roman Empire. At present, Spa is a major European spa and the venue of a renowned theatre festival, highly valued in Europe (Rubin et al. 2000 p. 114). Similar events are organised at other destinations connected with spa medicine in Europe. Another example is Baden–Baden in

Germany, a well-known spa where the Baden–Baden Easter Festival is held regularly.

The next period in the development of festivals came in the twentieth century, when all the civilisational factors which contributed to the growth of these events in the nineteenth century were intensified. At the time of the post-war revival, the fordist economy developed, based on the mass production of consumption goods. Cheap, mass production satisfied the growing social needs. As a result, the incomes increased, states gained more money from taxes, and new investments were made. Other changes included constructing common social systems, establishing workers' rights and introducing paid holidays. The middle class was formed, consisting of educated professionals, managers and owners of small- and medium-sized enterprises. Those economic and social advancements increased the financial surplus at households and the amount of available leisure time. As a result, societies in developed countries started to take more advantage of culture or tourism resources.

The 1970s were the time when the economy based on the fordist model was in crisis and a post-Fordist model of development was created. It was induced by the processes of deindustrialisation and globalisation as well as by the appearance of new technologies (the inventions of the third industrial revolution). Post-Fordism involved a departure from the economy based on mass production, large employment in traditional industries and a crisis of welfare state. Modern industries (high technology, IT), qualifications, creativity and workers' flexibility started to play a growing economic and social role. The economy entered the post-industrial phase, in which industry played an important role, but it was not so crucial any more. For instance, as regards employment, production was replaced by services, also those culture-related. Services, in turn, started to play an increasingly important role, also as regards generating incomes and taxes. A new group emerged, called the creative class (see works by Florida 2002a, b), and its growing role in the socio-economic development was noticed (see Amin 1994; Korec 2007).

As a consequence of those processes, the role of festivals started to grow, which led to the so-called festival boom in the second half of the twentieth century. Festival events of different kinds and sizes started to be organised in Europe, the USA and Australia. In some regions, their number reached hundreds or thousands annually. The festival boom was caused by many economic, civilisational and social factors, which emerged after the Second World War. Festivals became then an element of building the post-industrial economy. On the one hand, they became a valuable product which was a component of the cultural economy—a section of modern service economy, which generates and offers economically and socially valuable, culture-based products for sale. They became popular in the twentieth century and twenty-first century, during the time of the cultural turn, noticeable, e.g. in economy and science. The cultural economy includes products such as films, music, TV programmes, computer games, advertising, fashion, theatre performances, concerts, entertainment events and festivals. The growing affluence of societies and the growing interest in services satisfying higher and also culture-related needs formed the basis for the cultural economy development. Various authors also use the expression "cultural industries", referring to the components of the cultural economy. According to Cudny (2014c), "cultural industries deal primarily with symbolic goods, whose primary economic value derives from their cultural value. There are the 'typical' cultural industries (broadcast media, film, publishing, recorded music, design, architecture, new media), and the 'traditional arts' (visual art, crafts, theatre, musical theatre, concerts and performance, literature, museums and galleries). Cultural industries generate certain values, like income, taxes and jobs, and form the cultural economy" (see also Scott 2010; Gibson and Kong 2005). It can be seen that this type of activity may also include activity related to the organisation of festivals.

Other important factors related to the civilisational changes taking place in contemporary rich societies are the phenomena of so-called play ethic and experience societies. Such societies are also called hyperfestive societies or society of the spectacle (Davies 2015, p. 14). According to Rifkin (2000), even today people tend to move from work ethic to play ethic. This civilisational change is perceived in a similar way by Schulze (1992), who refers to contemporary societies as experience societies, whose members are looking for increasingly stimulating experiences. Such experiences could be understood as lifetime experience or one-time extraordinary experiences. Both could be developed through participation in such activities such as travelling, discovering new places and cultures, having fun or participating in exciting events, such as festivals. In Rifkin's opinion (2000), the further civilisation advances, the less time work takes in our lives, and more important are such experiences.

Experiences play a growing role in contemporary tourism as well. They put tourists in a pleasant state of arousal and satisfaction. Arousal, which is a particular state of mind, is connected with specific experiences, e.g. during a tourist trip. This function of experiences was noticed by the tourism industry, and today is the basis for the development of many tourist products. There is an opinion in psychology that feeling satisfaction is one of the major factors motivating people for action. The feeling of satisfaction may be achieved due to participation in different kinds of experiences containing exciting elements, such as risk, novelty or thrill. Risky experiences are among the currently very popular forms of adventure tourism. Among the tourist experiences offering novelty and thrill, the particularly significant products are those based on intellectual activity, such as different forms of art, literature, theatre and music (Anderson 2007, see also: Bryson 2007). According to Stasiak (2013), the turn of the twentieth century and twenty-first century was a period of a rapid development of the experience economy, in which the basic commodities are not only specific products, but also customers' emotions and experiences. In this sense, tourism has actually always been a kind of the experience industry. However, in recent decades, more pressure has been put on conscious creation of tourist products involving intense experiences. At present, multiplying and strengthening tourists' impressions are purposeful and intensive.

According to Quinn (2005, p. 937), "the search for experiential holidays supports the widespread orientation towards a greater consumption of cultural goods and experiences, including festivals". Also Getz (2008) claims that rapid development of the events' sector, including festivals, is a derivative of experience economy development. He believes that unusual experiences are the core of the product (different kinds of events). It is also the most important as regards attracting tourists to a given destination, also as a part of festival tourism.

Moreover, the global economy experienced rapid development of social mobility, especially after the Second World War. The contemporary world, where people move and communicate over long distances extremely fast, effectively and comfortably, can be defined as a mobile world (Urry 2007). A part of this extraordinary human mobility is tourism, which became a mass phenomenon after the Second World War (Urry 2002). A part of the whole phenomenon of tourism is the recently growing festival tourism (see Cudny 2013).

It should be stressed that culture and entertainment, as well as the broadly understood urban tourism, have become some of the most important urban functions. The development of the post-Fordist economy and services and the growth of mobility and tourism are naturally reflected in cities. The urban structure has changed, industry is not the leading activity any more, and modern services have developed (Pratt 2009). There appeared megacities, occupying the top position in the hierarchy of cities. Similar transformations concern other, smaller cities in the world, though on a more modest scale. The role of a contemporary city in the global network of cities is determined by elements related to culture and tourism, including festivals (Hall 1997). What is more, many cities and regions have started to use festivals for promotional purposes. Development policies called culture—led regeneration strategies, as well as event—or festival—led regeneration strategies, based on festivals, have been implemented. They concern cities which lost their economic base as a result of deindustrialisation. The way chosen to improve their economic situation and to promote them better is the development of cultural and entertainment services, including festivals.

The origins of the festival boom in the second half of the twentieth century were briefly presented by Cudny (2014b). He believes that the massive development of festivals at that time was caused by five main groups of factors. The first of them consisted of civilisational and new socio-psychological factors, first of all the greater amount of free time in contemporary societies. People gained more leisure time generally (the time off work, the time before finishing education in the case of young people and the retirement period in the case of elderly people), weekly (free weekends) and annually (paid holidays in the case of working people). There is no doubt that satisfying basic needs, which became possible in developed societies, generate needs of higher order, including those which are culture-related (also the need to attend festivals).

The second important factor given by Cudny (2014b) is the development of tourism, including the festival tourism. This is caused by the increasing average income as well as growing surplus of free time or development of average level of education. However, one of the most important factors of the development of tourism including festival tourism is increasing mobility. This human mobility is rising high in the second half of the twentieth century. This phenomenon was caused by civilisation development, due to growing income and increasing amount of free time. As a result, the members of developed societies started to move around the world in search for experiences. Very important impulse for the growing human mobility was also the phenomenon of globalisation. We can see growing role of international relations and flow of people as well as products in global economy. This also influences the development of mobilities including tourism (business tourism). Thanks to globalisation, there are also simpler custom and visa procedures. Thus, people have simpler travel procedures. The growing mobility (in professional and leisure purposes) must be treated as one of the most important factors triggering also the festival boom (for details about human mobility including tourism see Urry 2002, 2007).

The third group of factors is connected with a considerable increase in the average income in the twentieth century and with the possibility to spend its surplus, e.g. on culture. Moreover, the development of businesses, their growing incomes and searching for new means of promotion contributed to the development of festivals. They receive an increasing financial and organisational support from sponsors. The structure of festival management is improving as well, due to the development of research centres educating specialists in this field.

The last two groups of factors presented by Cudny (2014b) are also significant: the political transformations in former European communist countries and in China and the technological advancements and the birth of pro-ecological attitudes. Those were closely connected with the introduction of the freedom of assembly, opinions and attitudes. Those liberties reappeared in post-communist countries in the late twentieth century and contributed to the development of festivals there. Moreover, some modern festivals are based on pro-ecological ideas, currently very popular (e.g. healthy food festivals or film festivals concerning environmental protection.).

2.3 Types of Festivals

There are many different types of festival events. Some of them attract visitors from all over the world. Others are small street or estate festivals. There are one-off and regular events. Their themes vary, from art (opera, theatre) festivals to such strange events as, e.g. the festival of kites. Some people claim that meetings with family and friends, e.g. to celebrate someone's birthday, are also festivals. There are several works describing the variety of festivals worldwide which could be interesting for the reader (see Cooper 1995; Chanchreek and Jain 2007). Scientific literature contains many works presenting more or less formal typologies of festivals. However, there are few concise works which would summarise the research achievements in this respect.

Cudny (2014b) presents such a summary of the types of currently organised festivals. In this book, the analysis concerns a modified proposition published in the work cited above. The typologies of festivals are mostly based on the following criteria:

1. Attitude to religion
2. Festival venues

3. Social class structure, division of power and social roles
4. Important moments in personal life
5. Season
6. Scale and importance
7. Repeatability
8. Form of organising and financing the event
9. Structure of the festival guests and visitors
10. Theme
11. The historical situation and the geographic situation of special events.

As regards the first typology criterion (attitude to religion), we may divide festivals into secular and sacral, i.e. religion-related or referring to religion (see also: Falassi 1987). In the case of religious festivals, they may be further subdivided into different types of festivals depending on the organiser. For instance, there are festivals related to religion but organised by secular people somehow involved in religion, like the music festival entitled Rock for Trinity, organised in Łódź (Poland) by a Catholic association The Lord is great. On the other hand, there are festivals organised by churches, religious unions and clergymen, such as the Festival of Christian Culture, co-organised by the Religious Organisation of the Creative Milieu (a church organisation) in Łódź. There are also other traditional Christian festivals organised by the church.

Very good example of that kind of situation is Poland, where those festive events were organised even in communist times (1945–1989) and flourished after the fall of communism in 1989. We may give here the examples of several large festive events such as Feast of Corpus Christi, The Glorious Mystery of the Passion of Christ and Cavalcade of the Three Kings. The Feast of Corpus Christi Procession commemorates the Last Supper and the consecration of bread and wine into the body and blood of Jesus Christ. In Poland, it is celebrated on a Thursday always 60 days after Easter. This is public procession which takes place in the parishes, after the Holy Mass. It is conducted by the local priests, and during the procession, the most important sacred places in the parish are visited, including street crosses and shrines. The procession stops at field altars, where the fragments of the Bible are read.

Very important festive event connected with Catholic Easter is The Mystery of the Passion of Christ. The Mystery includes events commemorating the death and resurrection of Jesus Christ. A particularly important part of the Mystery is the Stations of the Cross. It involves a symbolic reconstruction of the road for Jesus Christ to death. This kind of celebration connected with parades and ritual reconstruction of the Stations of the Cross takes place in many Polish pilgrimage destinations. One of them is large sanctuary known as Kalwaria Zebrzydowska, where such mysteries are organised since 1806. Today, mysteries in Kalwaria Zebrzydowska are the target of annual arrivals for thousands of pilgrims travelling from Poland as well as from abroad.

Another example of quite new tradition, which roused after 1989, is the Cavalcade of the Three Kings (Epiphany). As Przybylska (2015, p. 176) wrote "In

2014, it was estimated that as many as 630 000 people took part in cavalcades in 177 towns and villages in Poland. Additionally, since 2013 the Polish Foundation has been the patron of some cavalcades abroad. In 2013, there were 6 events in Spain, Ukraine, the UK and in the Central African Republic, and a year later 10 processions in the USA, the UK, Ruanda, Ukraine, Italy and Germany".

Good example of religious festival came in last year from China, where after the political shift in the 1990s, these events started to be supported and revived with the help of regional authorities. They are treated not only as religious events, which help to cultivate the religious traditions within the society (as it is in Poland), but this is rather as tourist generators. The central government and regional authorities perceive those events as tools for regional development on the basis of festival tourism. However, on the other side such festivals as the Ghost Festival or Buddhist Festival in Wutaishan, China still have strong religious element. Thus, we should ascribe them into the type of religious festivals (see Teiser 1988; Ryan and Gu 2010). We have similar situation in the case of Hindu festivals in India like Maha Kumbh Mela (see Davies 2015) or different Shinto-related festivals in Japan—like Gion Festival (see Roemer 2010).

As for the next criterion, i.e. the festival venue, there are many variants. We may distinguish between festivals organised in the country and in cities and festivals held in the open air (e.g. in a park, amphitheatre) and in roofed facilities (e.g. halls, culture clubs, galleries). Another division may be that into seaside, mountain and other festivals.

The third criterion is based on the social class structure, division of power and social roles and refers to the classification presented by Falassi (1987), who distinguished several types of festivals: those organised by the people for the people (e.g. small neighbourhood festivals), by the elite for itself (e.g. high culture festivals in which elite representatives take part, such as Salzburg Festival or Bayreuth Festival), by the people for the elites, by the elite for the people (e.g. all kinds of military parades, such as the one in Paris organised on the National Holiday of France on 14th of July) and by the people against the elite. An example of festivals organised by the elite for the people could be find during the communist period in Poland and Soviet Union. A good example is so-called Labour Day celebrated on the 1st of May. This is a holiday developed to honour the workers' protests organised in 1886 in Chicago. In Poland, as well as in many other communist countries, it was a state holiday since 1950. Its celebrations (during the communist era) were combined with a parades and artistic performances. The parades involved different organisations, authorities and politicians at different levels. After 1989, these festival is not so pompously celebrated, but still in many Polish locations, we have parades or marches organised by social organisations or trade unions.

Another example of such communist state holiday connected with festive events such as parades and marches is the Anniversary of the October Revolution. It was celebrated on November 7 to commemorate the Great Bolshevik Revolution in Russia (1917). In Poland, it was the state holiday after the World War II. During these days, official academies were organised in schools, public institutions and factories. Besides, the military parades and ceremonies organised at the monuments of Polish–Soviet friendship were part of official ceremonies. This was also very important state holiday in Soviet Union, where large military parades were organised, by the communist regime, on the Red Square in Moscow to commemorate the anniversary of the Great Bolshevik Revolution.

An example of the festive event organised by the people against the elite was Woodstock 1969. It was a festival of rebellion against the lifestyle, politics and social relations of those times. A similar role was played by the Polish Jarocin Festival, which was an expression of the youth's rejection of communism and its elites. Those festivals were organised for the first time on the basis of counter culture activities, which appeared in the 1960s and 1970s. Young generation from a kind of rebellion against the social norms, and organisation of festivals was a part of this rebellion (see Johansson and Toraldo 2015).

Considering important moments in personal life, we may distinguish events or festivities referring to birth, marriage or death (see also: Duvignaud 1989). It is certainly possible to quote examples of events connected with the First Holy Communion, so popular in Poland in recent years or the Bar Mitzvah celebration, during which a Jewish boy officially comes of age.

As regards the year season criterion, festivals are usually divided into those held in winter, spring, summer and autumn (see Aldskogius 1993; Ryan et al. 1998; Visser 2005). Naturally, events organised in cold periods are usually organised in buildings. The work which seems interesting as regards this criterion is that by Cudny and Rouba (2011), presenting the distribution of Łódź theatre and multicultural festivals in time (Fig. 2.3).

As for the scale and importance criterion, Cudny and Rouba (2011) presented a concise classification of festivals into three categories. The authors distinguished among large festivals (well known and commented on in the media, mostly long-lasting, visited by large numbers of people [1000 or more], with many performances, international), medium-scale (less known and commented on in the media, shorter, with fewer visitors [less than 1000], mainly international or national) and small (not well known, with fewer visitors [less than 1000] including events devoted to the limited audience like the fans of stage magic and illusion art, artists [actors] and amateurs). A similar typology can be found in the work by Jago and Shaw (1998), who divided events into minor, major, hallmark and mega events. A classification of events according to their scale and importance was also suggested by Getz (2008), who divided them into local, regional, periodic hallmark events and occasional mega events. Those classifications may well be used for the division of festivals.

As for the criterion of repeatability, we deal with one-off and repeatable, regular events. When analysing their organisational form and the ways of financing them, we may speak of festivals organised spontaneously. This method of organisation may prove itself in the case of small events created by local communities, e.g. the inhabitants of an estate. In the case of large festivals, we have a situation when the event is organised by an institution, e.g. an association, foundation or city authorities.

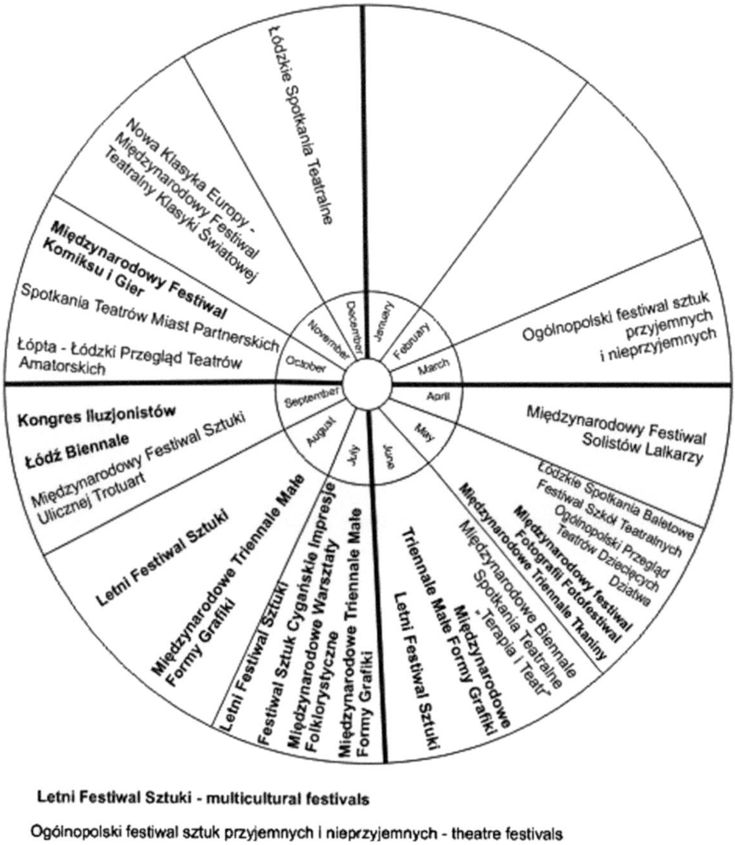

Fig. 2.3 The distribution of Łódź theatre and multicultural festivals in time. *Source* Cudny and Rouba (2011, p. 13)

If we consider ways of financing, there are festivals financed from private sources, e.g. by sponsors, or means paid to the account of a foundation or association organising the event. There are also festivals financed from public sources, e.g. the city, government or European funds. Another type of events includes those financed with mixed means—public and private.

As for the next criterion, i.e. the structure of festival guests and spectators, we may talk about several types of events. There are festivals whose guests are, e.g., artists or travellers presenting their achievements, invited by the organisers from the area of a given country (national festivals), from the given country and abroad or only from abroad (international festivals). A similar division can be made when analysing the structure of the event audience, which is more difficult because it requires more advanced studies (see Cudny and Stanik 2013; Cudny and Ogórek 2014). Moreover, we may talk about community-based festivals, with guests coming from the local community, e.g. the housing estate or district. We may also distinguish multicultural festivals, which present a multicultural and multiethnic mixture of themes. They are visited by artists and guests from different nations, minorities and countries. Apart from that, a separate type may be gay and lesbian festivals (see Lughes 2006; Gorman-Murray et al. 2008).

The next criterion of festival typology concerning their themes shows an exceptional variety. We may distinguish many types, such as art festivals (theatre, film, book, sculpture, painting, comprising different arts), folk festivals, festivals of food and wine, fashion events, events concerning adventure and tourism, those dedicated to sexual minorities—gay and lesbian festivals, festivals of science and technology, multicultural festivals and events dedicated to the ethnic minority inhabiting a given region and many other. For example, on the basis of his research in South Africa, Visser (2005) distinguished among four types of festivals. "The first cluster relates to broadly-speaking agriculture-related festivals including general agricultural produce, wine and specialised foods (…) The second cluster of festivals focus on the arts, both performing, and visual, and more generally combinations thereof (…). A notable third festival cluster (tagged "other" in this study) focuses on combinations of both, arts and agriculture, as well as including general trade festivals, combined with entertainment, often targeting family audiences. (…) A fourth focus relates to the festivals dealing with cultural or ethnic group-identities, such as Zulu history, or early colonial settlers. This focus also includes identity groupings, such as the gay community, who have no fewer than four annual festivals to choose from (Visser 2005, p. 169)". As for individual themes, this work presents the following types of festivals: art, art and music, dance, drama, film, general arts, key arts, jazz, music, identity based, tourism, sport, special interest, agriculture, food, food and wine, cheese and wine, wine, etc.

A different kind of festival is so-called boutique festival describe among others by Johansson and Toraldo (2015). Those authors argue that these are small-scale highly participative elegant intimate events, mostly devoted to different kinds of art (thus those festivals might be treated as a kind of art festivals). This kind of event allows its participants to engage into different social contacts and offer individual intellectual and aesthetic satisfaction. Taking part in such a festival is also a way of escape from (cultural) mainstream and engagement in counterculture and anti-corporate sensibilities.

The final criterion of festival typologies regards the historical situation and the geographic situation of festival events and was described in the publication by Ma and Lew (2012). These authors distinguished four types of festivals. First are local heritage festivals which are deeply rooted in local historical heritage; thus, those events are shaped by geography (local events) and time (they refer to history). Second type are national heritage festivals those are also connected with the history of a nation/country, but not strictly linked to specific local area. Such events operate on a nation/country level. Third type is strictly based on the geographic dimension; here, we have place-specific and place-non-specific festivals. First kind is connected with local space, and such events exist with reference to specific geographical and heritage context. The second type is not connected to specific place and could be organised in almost any location. The last type is based on the historical dimension;

here, we deal with traditional and modern festivals. Traditional festivals refer to the religious customs, agricultural practices, and the reenactment of historical events. Those events reflect a particular locale history and traditional culture. Modern festivals refer to the contemporary life, entertainment and recreation, and they are often tourist destinations. These events tend to be postmodern centres of consumption, where media appearances play significant role (Ma and Lew 2012).

Study Tips:

- What are the features differentiating festivals from other types of events?
- Describe the history of festivals from antiquity to the contemporary times.
- Give the main factors of festival development in the twentieth century.
- Discuss the criteria of classifying festivals and give their main types.

References

Aldskogius H (1993) Festivals and meets. The place of music in "Summer Sweden". Geogr Ann Ser B Hum Geogr 75(2): 55–72
Alexander JC (ed) (1988) Durkheimian sociology: cultural studies. Cambridge University Press, Cambridge
Amin A (ed) (1994) Post Fordism: a reader. Blackwell Publishers Ltd., Oxford
Anderson TD (2007) The tourist in the experience economy. Scand J Hosp Tour 7(1):46–58
Arcodia C, Robb DA (2000) A taxonomy of event management terms. In: Allen J, Harris R, Jago LK, AJ Veal (eds) Events beyond 2000: setting the agenda proceedings of conference on event evaluation, research and education, Sydney July 2000, pp 154–160, Australian Centre of Event Management School of Leisure, Sport and Tourism University of Technology, Sydney
Ashby C (1999) Classical Greek theatre: new views of an old subject. University of Iowa Press, Iowa City
Barnes TJ (2001) Retheorizing economic geography: from the quantitative revolution to the "cultural turn". Ann Assoc Am Geogr 91(3):546–565
Barnett C (1998) The cultural turn: fashion or progress in human geography? Antipode 30(4): 379-394
Brennan-Horley C, Connell J, Gibson C (2007) The Parkes Elvis revival festival: economic development and contested place identities in rural Australia. Geogr Res 45(1):71–84
Bryson JR (2007) Arts, dance, cultural infrastructure, and city regeneration: knowledge, audience development, networks, and conventions, and the relocation of a Royal Ballet company from London to Birmingham. Nor J Geogr 61(3):98–110
Caillois R (2001) Man and the sacred. University of Illinois Press, Chicago
Chanchreek KL, Jain MK (2007) Encyclopaedia of great festivals. Shree Publishers & Distributors, New Delhi
Cooper JC (1995) Dictionary of festivals. Thorsons, London
Corneanu S (2011) Regimens of the mind: Boyle, Locke, and the early modern cultura animi tradition. The University of Chicago Press, Chicago
Cruikshank J (1997) Negotiating with narrative establishing cultural identity at the Youkon International Storytelling Festival. Am Anthropol 99(1):56–69

Cudny W (2013) Festival tourism—the concept, key functions and dysfunctions in the context of tourism geography studies. Geografický časopis 6(2):105–118
Cudny W (2014a) Festivals as a subject for geographical research. Geografisk Tidsskrift-Danish J Geogr 114(2):132–142
Cudny W (2014b) The phenomenon of festivals—their origins, evolution and classifications. Anthropos 109(2):640–656
Cudny W (2014c) The influence of the "Komisarz Alex" TV series on the development of Łódź (Poland) in the eyes of city inhabitants. Morav Geogr Rep 22(1):33–43
Cudny W, Ogórek P (2014) Segmentations and motivations of the attendees' of the Mediaschool Festival in Łódź, Poland. Bull Geogr Socio-econ Ser 24:41–56
Cudny W, Rouba R (2011) Theatre and multicultural festivals in Lodz as a free time management factor among the inhabitants and tourists in the post-industrial city. Acta Geogr Univ Comenianae 55(1):3–22
Cudny W, Gosik B, Rouba R (2011) Festiwal kultury chrześcijańskiej w Łodzi jako przykład produktu turystycznego – wydarzenie. Ekonomiczne Problemy Usług 66(648):89–100
Cudny W, Stanik E (2013) "Man in danger" Media Festival in Lodz—the structure and motivations of the festival visitors. Ann UMCS Sect B (Geographia, Geologia, Mineralogia et Petrographia) LXVIII(1):69–90
Davies WKD (2015) Festive cities: multi-dimensional perspectives. In: Davies WKD (ed) Theme cities: solutions for urban problems. Springer, Chaam, pp 533–561
De Valck M (2007) Film festivals: from European geopolitics to global cinephilia. Amsterdam University Press, Amsterdam
Duvignaud J (1976) Festivals: a sociological approach. Cultures III(1):13–25
Duvignaud J (1989) The festive spirit. The UNESCO Courier XII:11–17
Eco U (1984) The frames of comic freedom. In: Eco U, Ivanov VV, Rector M, Sebeok TA (eds) Carnival, Walter de Gruyter &Co, Berlin, pp 1–10
Falassi A (1980) Folklore by the Fireside: text and context of the Tuscan Veglia. University of Texas Press, Austin
Falassi A (1987) Festival: definition and morphology. In: Falassi A (ed) Time out of time. University of New Mexico Press, Albuquerque, pp 1–10
Florida R (2002a) Bohemia and economic geography. J Econ Geogr 2(1):55–71
Florida R (2002b) The rise of the creative class. Basic Books, New York
Gay P, Pryke M (eds) (2002) Cultural economy: cultural analysis and commercial life. Sage Publications Ltd., London
Getz D (1991) Festivals, special events, and tourism. Van Nostrand Reihold, New York
Getz D (2005) Event management and event tourism. Cognizant Communication, Elmsford
Getz D (2008) Event tourism: definition, evolution, and research. Tour Manag 29(3):403–428
Getz D (2010) The nature and scope of festival studies research. Int J Event Manag Res 5(1):1–47
Getz D (2012) Event studies. Theory, research and policy for planned events. Routledge, London and New York
Gibson C, Connell J (eds) (2011) Festival places: revitalising rural Australia. Channel View Publications, Bristol
Gibson C, Kong L (2005) Cultural economy: a critical review. Prog Hum Geogr 29(5):541–561
Gibson C, Stewart A (2009) Reinventing rural places. The extent and impact of festivals in rural and regional Australia. Research Report. Australian Research Council, University of Wollongong. http://www.uow.edu.au/content/groups/public/@web/@sci/@eesc/documents/doc/uow060229.pdf. Accessed 01 May 2015
Gibson C, Connel J, Waitt G, Walmsley J (2011) Extent and significance of rural festivals. In: Gibson C, Connell J (eds) Festival places: revitalising rural Australia. Bristol, Tnawanda, Channel View Publications, pp 3–24
Gierczak B (2011) The history of tourist transport after the modern industrial revolution. Pol J Sport Tour 18(4):275–281
Goldblatt JJ (1990a) Special events: the art and science of celebration. Van Nostrand Reihold, New York

Goldblatt J (1990b) Special events: the art and science of celebration. Wiley, New Jersey
Goldblatt J (2000) A future for event management: the analysis of major trends impacting the emerging profession. In: Allen J, Harris R, Jago LK, Veal AJ (eds) Events beyond 2000: setting the agenda proceedings of conference on event evaluation, research and education, Sydney July 2000. Australian Centre of Event Management School of Leisure, Sport and Tourism University of Technology, Sydney, pp 1–8
Gorman-Murray A, Waitt G, Gibson C (2008) A Queer Country? A case study of the politics of gay/lesbian belonging in an Australian country town. Aust Geogr (2): 171–191
Grzesiak D (2010) Mysterion: misteria w Eleusis. http://www.histurion.pl/historia/starozytnosc/grecja/mysterion__misteria_w_eleusis.html. Accessed 31 Oct 2014
Hall CM (1989) The definition and analysis of hallmark tourist events. GeoJournal 19(3):263–268
Hall CM, Sharples L (2008) Food and wine festivals and events around the world: development. Management and Markets, Butterworth–Heinemann, Burlington
Hall P (1997) Modelling the post industrial city. Futures 29(415):311–322
Hauptfleisch T (2007) Festivals as eventifying systems. In: Hauptfleisch T, Lev-Aladgem S, Martin J, Sauter E, Schoenmakers H (eds) Festivalising! Theatrical events, politics and culture. Rodopi B.V., Amsterdam–New York, pp 39–47
Hawkins H (2013) Geography and art. An expanding field: site, the body and practice. Prog Hum Geogr 37(1):52–71
Isar RF (1976) Culture and the arts festival of the twentieth century. Cultures 3:125–45
Jacquot J (ed) (1956–1975) Les fêtes de la Renaissance, 3 vols, Editions du Centre national de la recherche scientifique, Paris
Jacobs MD, Weiss Hanrahan N (2005) The Blackwell companion to the sociology of culture. Blackwell Publishing Ltd., Malden
Jago L, Shaw R (1998) Special events: a conceptual and definition framework. Festival Manag Event Tour 5(2):21–32
Janiskee B (1980) South Carolina's harvest festivals: rural delights for day tripping urbanites. J Cult Geogr 1(1):96–104
Janiskee R (1991) Rural festivals in South Carolina. J Cult Geogr 11(2):31–43
Jędryka B (2012) Famous artists and the tourist assets of Krynica Zdrój associated with them. In: Michalski T, Radczenko A (eds) Global, continental, national and regional conditions of local development. Publishing House Adndu, Gdańsk-Kharkiv, pp 79–89
Johansson M, Toraldo ML (2015) 'From mosh pit to posh pit': festival imagery in the context of the boutique festival, culture and organization. Published Online. doi:10.1080/14759551.2015.1032287
Kelley DR (1996) The old cultural history. Hist Hum Sci 9(3):101–126
Kim S (2015) Understanding the historical and geographical contexts of food festival tourism development: the case of the Tatebayashi Noodle Grand Prix in Japan. Tour Plan Dev 12 (4):433–446
Klein RG, Blake E (2002) The dawn of human culture. Wiley, New York
Korec P (2007) Vplyv post-industriálneho štádia vývoja spločnosti a globalizácie na regionálny vývoj Slovenska. Geographia Cassoviensis I:75–80
Lughes HL (2006) Gay and lesbian festivals: tourism in the change from politics to party. In: Picard D, Robinson M (eds) Festivals, tourism and social change. Remaking worlds, Channel View Publications, Clevedon, pp 238–254
Ma L, Lew AA (2012) Historical and geographical context in festival tourism development. J Herit Tour 7(1):13–31
Muir E (1981) Civic ritual in Renaissance Venice. Princeton University Press, Princeton
Osnes B (2010) The Shadow Puppet Theatre of Malaysia: a study of Wayang Kulit with performance scripts and puppet designs. McFarland & Company Inc., Jefferson NC
Piette A (1992) Play, reality, and fiction: toward a theoretical and methodological approach to the festival framework. Qual Sociol 15(1):37–52
Pratt AC (2009) Urban regeneration: from the arts "feel good" factor to the cultural economy: a case study of Hoxton, London. Urban Stud 46(5–6):1041–1061

Przybylska L (2015) Between sacralization and festivalization of public spaces: a case study of the Cavalcade of the Three Kings in Poland. Bull Geogr Socio-econ Ser 27:171–180
Quinn B (2005) Art festivals and the city. Urban Res 42:927–943
Quinn B (2009) Festivals, events and tourism. In: Jamal T, Robinson M (eds) The SAGE handbook of tourism studies. Sage Publications Ltd., London, pp 483–503
Radziejewski P (2006) Saturnalia i ich związek z Bożym Narodzeniem. http://www.histurion.pl/historia/starozytnosc/rzym/saturnalia_i_ich_zwiazek_z_bozym_narodzeniem.html. Accessed 31 Oct 2014
Ray R, Sayer A (eds) (1999) Culture and economy after the cultural turn. Sage Publications Ltd., London
Rifkin J (2000) The age of access: the new culture of hypercapitalism, where all of life is a paid-for experience. J.P. Tarcher-Putnam, New York
Robertson R (2000) Globalization. Social theory and global culture. Sage Publications LTD., London
Roemer M (2007) Ritual participation and social support in a major Japanese festival. J Sci Study Relig 42(2):185–200
Roemer M (2010) Shinto festival involvement and sense of self in contemporary Japan. Japan Forum 22(3–4):491–512
Rubin D, Nagy P, Rouyer P (eds) (2000) World encyclopedia of contemporary theatre. Europe, Routledge, London
Ryan C, Gu H (2010) Constructionism and culture in research: understandings of the fourth Buddhist Festival, Wutaishan, China. Tour Manag 31(2):167–178
Ryan C, Smee A, Murphy S, Getz D (1998) New Zealand events. A temporal and regional analysis. Festival Manag Event Tour 5(1–2):71–83
Schulze G (1992) Die erlebnis gesellschaft. Kultursoziologie der gegenwart, Campus, Frankfurt
Scott AJ (2010) Cultural economy and the creative field of the city. Geogr Ann Ser B Hum Geogr 92(2):115–130
Stasiak A (2013) Produkt turystyczny w gospodarce doświadczeń. Turyzm 23(1):29–38
Strong R (1984) Art and power Renaissance festivals, 1450–1650. University of California Press, Berkeley
Strong R (2002) Feast: a history of grand eating. Harcourt, Orlando
Szczepański B (2012) The instrumental music of Wutaishan's buddhist monasteries: social and ritual contexts. Ashgate, Farnham
Tara Lunga M (2012) Major special events: an interpretative literature review. Manag Mark Challenges Knowl Soc 7(4):759–776
Teiser S (1988) The ghost festival in Medieval China. Princeton University Press, Princeton
Tondo JSF (2010) Popular religiosity and the transnational journey: inscribing Filipino identity in the Santo Niño fiesta in New Zealand. Asia Pac J Anthropol 11(3):219–244
Turner JH, Stets JE (2005) The sociology of emotions. Cambridge University Press, Cambridge
Tylor EB (1871) Primitive culture: researches into the development of mythology, philosophy, religion, art, and custom. John Murray, London
Urry J (2002) The tourist gaze. Sage Publications, Thousand Oaks, Calif
Urry J (2007) Mobilities. Polity Press, Cambridge, Malden
Visser G (2005) Let's be festive: exploratory notes on festival, tourism in South Africa. Urban Forum 16(2–3):155–175
Wasserman S, Faust K (1994) Social network analysis: methods and applications. Cambridge University Press, New York
Waterman S (1998) Carnivals for elites? The cultural politics of arts festivals. Prog Hum Geogr 22:54–74
Williams R (1976) Keywords: a vocabulary of culture and society. Oxford University Press, London–New York
Wilson N (ed) (2006) Encyclopedia of Ancient Greece. Routledge, New York
http://encyklopedia.pwn.pl/haslo/Dionizje;3892862.html—Online Encyclopedia of Polskie Wydawnictwo Naukowe. Accessed 20 Feb 2015

http://portalwiedzy.onet.pl/69310,,,,kultura,haslo.html—Online encyclopedia. Accessed 19 Oct 2014

http://www.britannica.com/EBchecked/topic/1567446/Grand-Tour—Online Encyclopedia Britannica, entry: grand tour. Accessed 23 Nov 2014

http://www.britannica.com/EBchecked/topic/203113/feast—Online Encyclopedia Britannica, entry: festival. Accessed 23 Nov 2011

http://www.britannica.com/EBchecked/topic/96363/Carnival—Online Encyclopedia Britannica, entry Carnival. Accessed 23 Nov 2014

http://www.histurion.pl/historia/starozytnosc/grecja/mysterion__misteria_w_eleusis.html—Website about the history of ancient Greece. Accessed 31 Oct 2014

http://www.oed.com/view/Entry/69567?redirectedFrom=festival+#eid. Onine Oxford English Dictionary. Accessed 3 Feb 2015

Chapter 3
Research Methodology in the Analysis of Festivals

Abstract In this chapter, the key theoretical and methodological aspects of a scientific festival analysis are discussed. The first part of the chapter presents the different aspects of research on festivals. The reader can find here the characteristic of how this issue is studied in different sciences such as event studies or geography, describing the origins and study aspects. This part of the chapter explains why scientists became interested in festivals in the second half of the twentieth century and describes in detail major research themes, explored mainly as a part of these sciences. The second part of the chapter is devoted to the methodological issues. Beginning of the chapter presents basic terms, such as method or methodology, and next, research methods used in the study of festivals are presented. Methods are divided into general and detailed, quantitative and qualitative, and field and laboratory methods. Another part of the second section of this chapter is devoted to a detailed description of the main types of research methods used in the analysis of festivals. Using numerous examples, quantitative and qualitative, and field and laboratory methods used for studying this phenomenon are characterised. This chapter is especially aimed at students researching different aspects of festival development and its various impacts.

Keywords Festival research themes · Festivals in event studies · Geography of festivals · Methodology of festival studies · Main research methods

3.1 Basic Research Themes

As it was mentioned in the previous chapter, festivals undergo analysis in many scientific disciplines. The increase in the scientific interest in festivals was caused by the growing social and economic role of culture and the growing interest in culture in general in the science world. As pointed out by Cudny (2014b), festivals were first studied mainly in sciences concerned with people and their culture, i.e. sociology or anthropology. The first works referring to festivals appeared at the turn of the nineteenth and twentieth century, but the boom of this kind of research came in the second half of the twentieth century, when the analysis of festivals developed as a

W. Cudny, *Festivalisation of Urban Spaces*, Springer Geography,
DOI 10.1007/978-3-319-31997-1_3

part of sociological and anthropological studies (see Pieper 1965; Duvignaud 1976, 1989; Isar 1976; Falassi 1987; Turner 1982; Piette 1992; Gotham 2005).

Furthermore, researchers representing other sciences wrote a number of general, theoretical papers, as well as other works devoted to specific issues, concerning the influence of festivals on culture, society or economy. We may quote here the example of human geography, where the first serious analyses of festivals appeared in as early as the 1980s (Quinn 2005a), but this research trend really flourished in the next decades. What is more, festivalisation started to be studied also in other sciences, such as economics or management. These studies concerned with the economic functions of festivals as well as their organisation and management.

Festivals are often treated as a type of events (which was shown in the previous chapter). Therefore, it is essential that the research trend called event studies should be discussed more broadly. Events were studied in sciences concerned with tourism management already in the 1970s (Quinn 2009). However, at the turn of the 1980s and 1990s, a new, separate scientific discipline, called event studies, started to develop. It referred to sociology, economics, management, tourism sciences and geography, and concerned all kinds of events. The first publications included those by Ritchie (1984), Hall (1989), Getz and Frisby (1988), Goldblatt (1990) and Getz (1989, 1991). The next years brought a systematic development of the event sector, due to a number of economic and social factors. According to Goldblatt (2000, p. 3), "(…) four changes: aging, technology, income and time have dramatically increased the demand for a wide variety of events, both in the U.S. and throughout the world". A huge increase in the number and variety of events in the second half of the twentieth century, particularly in the 1980s and 1990s, is referred to as a boom. With regard to festivals, we use the term "festival boom" (Cudny 2014b).

The development of events resulted in a growing need for research and education of professionals dealing with planning and organising of events. In response to this need, event studies integrated experts from many domains, practitioners and theoreticians concerned with the phenomenon of events. "Event studies is the academic field devoted to creating knowledge and theory about planned events. Every field requires a unique core phenomenon, and it is the study of all planned events that most obviously distinguishes event studies. This encompasses their planning and management, outcomes, the experience of events and meanings attached to them, and all the dynamic process shaping events and the reasons why people attend them" (Getz 2012, p. 4). Event studies are presently a very broad scientific discipline. Many universities offer courses educating specialists on event organisation. There are also many journals dedicated to event studies.

Events,[1] including festivals, started to be used for the promotion of countries, regions and cities as attractive, culturally rich areas (Quinn 2009). They also became an element attracting tourists and generating tourist traffic within a separate type of tourism, called event or festival tourism (see Robinson et al. 2004; Picard and Robinson 2006; Cudny 2013). As a result, they also became an economy

[1]The definition and typology of events were presented in Sect. 2.1.

development factor and started to be used in the strategies devised for cities, regions and whole countries. The variety of event types and their wide-ranging impacts is reflected in the multitude of research themes explored in the event studies.

Due to the fact that festivals are a significant research theme in event studies, it was decided to present the most important, well-established research approaches to festival analysis. In his work, Getz (2010, pp. 6–9) pointed to nine research issues in event studies, referring to the experience and meanings of festivals:

1. Political and sociocultural discourse on the meanings and effects of festivals: it refers to the political aspects of festivals connected with power relationships.
2. Authenticity and commodification: These involve studies on the authenticity of festival message and on its possible distortions caused, for example, by its commercialisation for the purposes of festival tourism.
3. Community, cultural, place identity and attachment: These regard the roles of festivals in creation and strengthening of cultural identity and their role in development of culture in certain places and groups.
4. *Communitas*, social cohesion and sociability: *communitas* refers to feelings of belongings as well as sharing among equals, as in pilgrimage or festival experiences. This research trend describes the existence and importance of *communitas* during festivals.
5. Festivity, liminality and the carnivalesque: liminality or the temporary state of being apart from the mundane is an enduring theme here.
6. Rites and rituals, religion: this theme discovers the connection between ritual and festivity, much of which has religious or spiritually symbolic significance.
7. Pilgrimage: festivals as destinations for religious or spiritual pilgrims.
8. Myths and symbols: research about the myths and symbols showed and used during different festivals.
9. Spectacle: festivals as spectacles—for example as cultural performances.

In another part of his work, Getz (2010) distinguishes three main discourses including festival analysis in event studies. The first one is the discourse on the roles, meanings and impacts of festivals in society and culture. The second one regards festival tourism, and the third—festival management. Modifying Getz' work (2010), Cudny et al. (2012) identified three research themes in analyses devoted to festivals, explored as a part of event studies (see Table 3.1):

1. Research theme: An analysis of the influence of festivals on culture and society. It is based on anthropological and sociological studies. This broadly represented research theme includes festivals as the place of cultivating intangible culture and as the space of promoting ideas, e.g. multiculturality. Moreover, the studies within this theme cover the festival visitors' motivations to attend the events and the segmentation of festival audience by their personal details (e.g. age, sex, family status, income). Other studied elements are the audience's satisfaction, the behaviours of event organisers, staff and volunteers, and the perception of events by the inhabitants of the host cities and regions. We may divide the main theme into subthemes concerning the motivations to attend festivals, the structure of the

audience (according to demographic and social features), the visitors' satisfaction, the behaviours of organisers and volunteers working at the festivals and the perception of events by the inhabitants of the host areas (see: Table 3.1).
2. Research theme: research on the influence of festivals on the economy, based on economics and regional management. These issues have an extremely significant practical dimension, because economic effects are often the main motivation to organise and finance a festival. In this case, it is established whether a given festival is a tourist asset, whether it attracts tourists or is an event oriented towards local inhabitants. When an event is a festival tourism objective, i.e. it is the main or at least additional aim of a given group's tourist trip, it brings large profits and generates jobs. The income is generated by the tourists' expenses on accommodation, petrol, entertainment and festival tickets. Moreover, festival events are an important tool for creating a positive image of the destination. Organising them increases the popularity of the area, shapes its positive brand and is a part of promotion strategies based on the development of culture and events. Due to such activities, cities, regions, or even whole countries attract more tourists, investors and migrants. This in turn contributes to their socio-economic development (see Table 3.1).
3. Research theme: works regarding proper management of festivals. They are publications concerning the practical organisation and management of this kind of events. The issues discussed in them regard planning events, financing them from public and private sources (sponsors), adjusting the event programme and venue to the visitors' expectations, and promoting events (see: Table 3.1).

Table 3.1 Selected scientific works regarding festivals, written as a part of event studies

Research theme	Major scientific works
Impact of festivals on culture and society	Mohr et al. (1993), Uysal et al. (1993), Saleh and Ryan (1993), Backman et al. (1995), David (1995), Formica and Uysal (1995), Gamson (1996), Schneider and Backman (1996), Crompton and McKay (1997), Grant and Paliwoda (1998), Krausse (1998), Nicholson and Pearce (2000), Delamare (2001), Delamere et al. (2001), Grant and Paliwoda (2001), Nicholson and Pearce (2001), Molloy (2002), Gursoy et al. (2003), Van Zyl and Botha (2004), Yuan et al. (2004), Lee et al. (2004, 2007, 2012), Bowen and Daniels (2005), Saayman and Saayman (2006), Wood and Rhodri (2006), Vinkle van and Backman (2009), Axelsen and Swan (2010), McDowall (2010), Buch et al. (2011), Abreu Novais and Arcodia (2013)
Studies on the impacts of festivals on the economy of cities and regions	Della Bitta et al. (1977), Long and Perdue (1990), Crompton and McKay (1994), Jago et al. (2003), Avraham (2014)
Problems of festival management	Getz and Frisby (1988), Hall (1992), Saleh and Wood (1998), Getz (2000), Long (2000)

Source Author's compilation

Another approach to the dimensionality of festivals as well as to their possible scientific analysis (based on these dimensions) has been proposed by Davies (2015). He presented a model describing the dimensionality of festive events, which comprises eight dimensions according to which festivals could be understood and assessed. It can be added that these dimensions may also be treated as some kind of festival research themes (Davies 2015, p. 544):

1. The ephemeral character of festivals. We deal here with festivals as temporary and spatially unstable phenomena. On the one hand, they last for a short time, e.g. a few days, and then disappear from the urban space. They may recur (regular festivals) but not necessarily (one-off, occasional festivals). On the other hand, festivals are often ephemeral events as regards the space they occupy. They often take up space normally used by other functions, e.g. transport (streets, squares), and disappear from these spaces right after they finish. In this case, scientific analysis may concern the seasonality and repeatability of events, or the way the urban space is used for the time of their duration.
2. The second dimension of festivals is their form. The focus point is whether an event is based on a single form of artistic expression (e.g. a concert, theatre performance, street parade) or several such forms (e.g. a festival combining various arts, such as music, dance and theatre). Moreover, as Davies emphasises (2015), the important thing is whether the form of the festival is authentic, i.e. based on authentic and living heritage, or whether it recreates some old local traditions for purely commercial purposes. In this case, the analysis of festivals may concern their types (depending on the form of presentation), as well as the authenticity of the message they convey.
3. Festival functions. It is the third dimension of festival events, each of which plays different roles in social and economic life. The author of the cited work differentiates among the following functions of festivals: celebration of cyclical (seasonal) events (such as harvest), religious beliefs, political functions, celebrations of traditions, cultural functions and lifestyle functions. In this respect, festival analysis seems to be a significant aim of scientific research. Moreover, it is impossible to understand festivals and festivalisation without referring to their various functions, i.e. the roles they play in the contemporary world.
4. Experiences as one of festival dimensions. As pointed out by Davies (2015), a festival is a powerful experience, which may take two different forms. The first involves experiencing an event as a happening evoking satisfaction, pleasure, developing one's knowledge and enriching human life. On the other hand, a festival is a transgressive experience, as it often involves breaking social and moral norms. It sometimes has a negative side, such as the lack of acceptance by the local community or some pathologies emerging during the event (e.g. riots, alcohol and drug abuse).
5. Exchanges as a festival feature. Since the beginning, exchange has been a part of festival events. In the past, during religious festivals or those devoted to agriculture, offerings were exchanged for the gods' graces. At present, exchange

usually has a different character, e.g. festival attendees exchange money for knowledge, good fun, sometimes also goods purchased during events (e.g. food or wine festivals). Currently, the economic role of festivals connected with financial exchange seems to be one of the most important practical dimensions of these events. The analysis of this kind of flow of money and the role played here by festivals is a major research theme.

6. Festival connectivities. This aspect of festivals is very important because they are places attracting people of similar interests, specialists in a given line of business or different kinds of enthusiasts. Festivals are also occasions for contacts between tourists and the local community. Thus, they are a place and time connecting people, a space of exchanging ideas, contact with art, etc. Festivals are places of constructive exchange but may also be areas of conflicts between festival visitors and the local inhabitants. These are important research issues, focusing on the problems of festivals as points of passage, human and cultural capital or social networks.
7. Festival constructiveness. It is extremely important from the practical point of view. This particular festival feature refers to the fact that their preparation requires taking a number of actions, cooperation of many specialists, a guarantee of subsidies, preparing the stage, artists, attracting visitors, etc. These activities are a part of constructing the festival and take place in three stages, i.e. before, during and after the event. Analyses of this festival dimension are extremely significant and are conducted as a part of event studies.
8. Impacts of festive events. "Like so many of the other dimensions (*described before*) it (*festival*) can result in alternative and often contradictory features, in this case gains or losses" (Davies 2015, p. 551). In this book, a separate subchapter is devoted to this issue (see: Sect. 4.2). It is about the impacts related to festival functions and observed in the host locations. There may be different types of impacts, e.g. an influence on the local community, culture, politics or, finally, economic effects, broadly described in the literature (e.g. in event studies). Currently, studying this kind of festival impacts, especially when divided into positive and negative ones, seems to be one of the most important research areas.

We should ask ourselves the question about the role of the elements of culture, including festivals, in geographical research, because geography has a long tradition of studying festivals also as regards their influence on cities. Festivals are thoroughly researched in geographical subdisciplines, such as cultural and tourism geography. Contemporary culture-related phenomena are an object of human geography studies. First of all, it must be stressed that culture is created by people who change natural environment through their civilisational (cultural) activity, adjusting it to their needs. They replace natural environment with geographical environment, whose increasingly important part is the sphere of human creation, i.e. anthroposphere (Gr. Atropos—human) (Wrona and Rek 1998, p. 16).

The geographical environment, which is created by people, undergoes geographical analysis. Geography as a science has been evolving for centuries;

however, it is certainly a discipline referring its studies to space (a chorological discipline). Regarding scientific definitions, Johnston (2007) defines geography after Hartshorne as a "(…) discipline that seeks to describe and interpret the variable character from place to place of the earth as the world of man".

The early geography typologies featured two basic disciplines—one related to natural elements (physical geography) and the other to human activity (human geography) (Hartshorne 1939). An additional basic geographical subdiscipline is the regional geography, which concerns regions as the analysis area (Leszczycki 1972). Nowadays, human geography includes many sub-disciplines, such as economic, social, political or cultural geography (Johnston et al. 1996, p. 260). Other subdisciplines of human geography are urban geography (Pacione 2005) and tourism geography (Kowalczyk 2000). Festival studies, discussed herein, may be included in many sub-disciplines of human geography, e.g. cultural, urban or tourism geography.

Festivals are a phenomenon which concerns space (cities, regions), and therefore, they are a study subject in geography (Waterman 1998). According to Cudny (2013), festivals are an element influencing tourism space studied in tourism geography. It is treated as a subspace of broadly understood geographical space (see Liszewski 1995). Cudny (2014a) also claims that festivals strongly influence geographical space; thus, they are studied in depth by geographers. "Festivals are of interest to geographers because they constitute one of many practices that humans have evolved in the process of making homes and carving out identities for themselves" (Quinn 2005b). Many festivals are an emanation of regional or local culture and express a close relationship between the cultural identity and a given space (Eckman 1999; Derrett 2003; Quinn 2003). Festivals also have a social and an economic impact on space as well as influence it through the facilities built for festival purposes.

As regards festival studies, one of the most important subdisciplines of human geography is cultural geography. We should ask ourselves why festivals are studied so extensively in this subdiscipline. As regards modern cultural geography, we may talk about the following main research dimensions (Anderson et al. 2003):

- Culture as distribution of things. Every culture produces various material artefacts, such as pieces of art, machines, buildings and clothes. An analysis of these elements and their diversification is one of the basic and oldest research issues in cultural geography.
- Culture as a way of life. Here, culture becomes the whole of human life, from its material dimensions to non-material ones, such as beliefs, customs or consumption behaviours.
- Culture as meaning. Here, tangible and intangible dimensions of culture are studied in terms of their significance. The things analysed include the importance of events from the point of view of the history of the site and the people who took part in them. Meanings may be different for different persons; they depend on history, customs and people inhabiting and visiting a given area.
- Culture as doing. In this case, culture is treated as a complex, interactive process of generating events and phenomena.

- Culture as power. Culture, its patterns, influence on people and space, and the ways it spreads are a kind of power. Some cultural patterns (present in the previous research dimensions) are more popular than others, often predominating in the contemporary world, while other ones are pushed to the margin of contemporariness. The analysis of the causes and results of such phenomena is also the object of study in cultural geography.

Festivals are a cultural phenomenon which has an influence on all the research aspects of cultural geography mentioned above. Thus, they are so often studied in this science. First, they leave behind physical objects, such as festival centres (buildings erected for festival purposes), sculptures (remaining after art festivals, e.g. festivals of sculpture) or murals (the result of street art festivals). They are a part of many people's lifestyle, e.g. young persons or lovers of a particular art. Apart from that, festivals are often connected with a given place or social group, expressing their history or belongings; they are an emanation of a community. Festivals are also an expression of culture creation, mainly as regards its intangible aspect, as well as an emanation of the power of culture. Thanks to them, it is possible to propagate ideas and cultural models; they have an influence on people, both individuals and societies.

In recent years, culture and its components have gained in importance as regards scientific studies, including geographical ones. The cultural turn in geographical debate, which was already mentioned before, began in the 1980s and developed in the 1990s. The role of culture and its different ways of shaping the world, influencing individual human decisions and developing strategies for cities or regions, started to be appreciated (Radclife 2006). Cultural geography developed a new approach called new cultural geography. According to Czepczyński (2007), it studies various manifestations of broadly understood culture, and its spatial aspects are used for reconstructing the identity of social groups. This new approach fully developed the main research themes quoted before, distinguished by Anderson et al. (2003), which obviously intensified festival studies.

Festivals are also often analysed as a phenomenon connected with urban geography. This geographical subdiscipline tries to explicate the distribution of cities as well as the socio-spatial similarities and differences among them and inside them (Pacione 2005). In this case, festivals are an element which modifies the social structure of cities by, for instance, influencing the social capital, culture development, or social attitudes. Festivals are also a medium that expresses the cultural identity, ideas and pursuits of many social groups inhabiting cities, e.g. minorities (e.g. multicultural festivals). Festivals also transform urban space, both temporarily, at the time of the event, and permanently, e.g. by building festival facilities. What is more, they also influence the urban economy by, for example, generating income from tourism or creating jobs in tourism services. The promotional role of festivals is equally important, as they promote and help to create a positive image of the host areas. Another significant function of festivals, though still insufficiently researched, is their use in the transformation of post-socialist cities. Recent years have brought a development of festival studies in post-socialist countries (see, among others:

Cudny 2013, 2014a, b; Bjeljac and Lovic 2011; Markova and Boruta 2012; Przybylska 2015). Before 1989, this discipline was nearly absent from geographical studies in those countries.

In many cities of the Western countries, we could observe regeneration strategies based on culture, art and entertainment, implemented among others in stagnating (e.g. post-industrial) cities (Binns 2005; Evans 2005; Smith and Fox 2007). In this type of strategies, events of various types, including festivals, were very frequently used. Event-based strategies are referred to as event-led regeneration strategies, and those which focus on festivals—festival-led regeneration strategies. Apart from many positive impacts, festivals also have some negative effects. Sometimes, they become a problem for cities, e.g. due to the conflicts growing around some events, which badly affect the image of urban regions (for an example of such conflicts see Cudny 2011).

Festivals have a substantial influence on tourism as well. First of all, they are tourist assets attracting tourists to a given destination (De Bres and Davis 2001; Quinn 2005a, b). Travelling to events of this type is treated as a part of cultural tourism. Such travels are also treated separately as so-called event tourism, which has a subtype called festival tourism (Cudny 2013). For this reason, festivals are also a research object in the broadly understood tourism geography. This research direction is defined as a geographical discipline, which "is concerned with the analysis of the forms and relations of spatial tourism phenomena, as well as related processes influencing the space. It examines:

- the usability of space in tourism, including its forms and seasonality,
- the impacts of tourism phenomena on the space,
- processes taking place in space, caused by tourism phenomena" (Warszyńska and Jackowski 1978, pp. 15–16).

Festivals are often the foundation of regional development and are also held in rural regions, where they are an element creating temporary work places, integrating local communities and enabling them to gain a number of organisational skills, attracting tourists and generating additional income (Brennah-Horley et al. 2007). Therefore, festival events are also studied as a part of rural geography or even regional geography. Examples of this kind of analyses can be found, e.g. in Australian geography (see Gibson and Connell 2011, 2012; Gibson and Davidson 2004; Gibson et al. 2010). Interesting work regarding the social and economic influence of festivals organised in rural environment in India was also published by Rao (2001). Also Ryan and Gu (2010) argue that the festival in rural areas of China is growing in recent years, and those events are a part of rural development strategies in China.

A detailed description of geographical festival studies was presented by Cudny (2014b, p. 136). Based on a thorough review of geographical literature, eight main research themes in the festival analysis in human geography were distinguished:

1. Cultural: regarding the analysis of festivals as places where culture-related phenomena occur, culture is promoted, created and consumed.

2. Social: devoted to the analysis of festival-goers, their origins, motivations to attend festivals and to the study of the host communities' attitude to the events and festival tourists.
3. Economic: an analysis of the economic impacts of festivals, including income from festival tourism, generating jobs, creating the image and regenerating crisis areas.
4. Time-space analysis: devoted to the spatial distribution of different types of festivals and their evolution.
5. Political: the subject of the analysis includes political and social ideas and their promotion during festivals.
6. Historical: research on the roles of festivals in the past.
7. General: combining several research themes.
8. Theoretical: devoted to different theoretical issues, such as the reasons for festival development, main roles of festivals and festival typologies.

This publication also presents the idea that the research on festivals is so advanced in human geography that a separate research trend, called the geography of festivals, could be established. Cudny (2014b, pp. 139–140) believes that "in the future, after standardising theoretical problems, such as the concept of festivals or typology, and establishing a cohesive research methodology, it will be possible to treat the geography of festivals as a separate sub-discipline of human geography. It must be remembered, however, that distinguishing such a sub-discipline requires further intensive theoretical, empirical and comparative research. An extremely important element in defining the separateness of geographical festival research (in relation to event studies) is referring these studies to geographical space. The geography of festivals should describe the changes, both physical (objective) and non-physical (subjective and abstract), produced by festivals in different dimensions of geographical space".

3.2 Research Methods Used in the Study of Festivals

Introduction

In order to be independent of and separate from other scientific disciplines, every science must apply clearly defined methods and have its own object of study. This book does not represent a separate science. However, we should point to the most important groups of methods and establish the basic techniques suitable for studying festivals, which will help the researchers who have just started to explore this phenomenon. Apart from that, the analysis of research methods will also be helpful for students, who are an important group of readers of this book. Moreover, one of the aims was to sort out some of the methodological issues. Currently, festivals are analysed in many various sciences, such as sociology, anthropology, management, event studies and geography. The studies include many different aspects of festivals and festivalisation. They cover a wide range of problems, such as the rules of organising and financing events, the visitors' structure and motivations, the perception of festivals by the inhabitants of host areas, the social and economic impacts of festivals and their role in creating the image of cities and regions. Such a wide range of research problems requires the use of a large variety of research methods.

At the beginning of this subchapter, it is important to define the basic terms connected with research methods. The first of them is methodology, which means "a science concerned with the structure of science and methods of scientific research" (Runge 2007, p. 18). We may talk here about universal methodology, which concerns research methods used in science in general, and specific methodology, comprising detailed methods. The latter concerns the research methods and the structure of individual scientific disciplines, so we may talk about the methodology of geography, sociology, anthropology, etc. (Cudny et al. 2011). A research method means a set of research techniques applied in order to describe a given problem, usually a key one in the given discipline. They are based on typical and recurring ways of collecting data and information, as well as analysing and interpreting it in order to obtain answers to the questions posed earlier. Research methods are then specific procedures used to reach the aims of study (Runge 2007, pp. 19–21).

According to Cudny et al. (2011), research methods may be divided into several categories, depending on the classification criteria. The first criterion could be that of universality. Here, we can distinguish between general methods of universal application and specific methods used in individual disciplines or groups of scientific disciplines (e.g. the group of social sciences). General methods may be used at every stage of a science development and practically in every discipline. They first of all regard thinking procedures and refer to the general, logical rules of drawing conclusions. In the case of specific methods, we cannot speak of such a large degree of universality, because they are adjusted to the individual object of study in a given science. For instance, it is difficult to apply survey methods, commonly used in social sciences, in physics or mathematics. On the other hand, specific methods are successfully used in related disciplines. A good example is social sciences, where a number of methods permeate from some disciplines to others. This process, sometimes referred to as science syncretism, has been proven in methodology in many ways. For example, survey methods (such as questionnaires), originally typical of sociology, are successfully used in human geography. Similarly, observation, which is a specific method typically applied in anthropology, has been effectively used in geography.

Another important classification of research methods is based on the research procedure. In this case, we distinguish between methods which make use of primary numerical data obtained, e.g. as a result of empirical studies and measurements, and a group of methods which are not based on the numerical values of phenomena. Thus, we have two types of research methods: quantitative and qualitative. The former are based on statistical data, often require using statistical indexes, computer programs, etc. (Veal 2006). They are mainly used in mathematical and natural sciences. However, the second half of the twentieth century brought a broader application of this type of methods in social sciences as well. We may quote here several examples, such as using statistical indexes in urban or population geography. In recent years, statistical analyses have been applied even in the study of literary texts.

Qualitative research methods are not based on direct acquisition of numerical data on the basis of a measurement or experiment. Here, we deal with an analysis of the perception of the world based on one's own experience or on the opinions of others, as well as on the significance ascribed to phenomena by people. In this case, the analysed elements are put through a special filter of individual perception by the researchers themselves or the individuals they study. In this kind of analysis, researchers' perception of the surrounding world is based on field study notes, interviews, conversations, photographs, recordings or memories (see Guest et al. 2013). The situation is typical of social studies, such as anthropology, sociology or pedagogy. The turn of the twentieth century brought the revival of qualitative methods in human geography, which to a large extent has always been based on this type of research. However, the 1970s and 1980s witnessed certain fascination with quantitative studies, based on direct measurements and the analysis of statistical data, using complicated statistical indexes. The late 1990s and the early twenty-first century were the time when the qualitative studies returned and methods borrowed from anthropology or sociology started to be applied more broadly in human geography (see Crang 2002, 2003, 2005).

The third classification which is important from the point of view of this book is the division of research methods depending on where the research is conducted. In this case, the methods fall into two groups. The first one includes field study methods, and the other one consists of laboratory research methods. Field study methods involve travelling to the place where a given phenomenon occurs. Using such methods is extremely important to obtain rational results. As regards festival studies, we should assume that we will not write about something which we have not observed personally. It is obvious that not all aspects or examples of a problem can be experienced by the researcher directly, but such experience, e.g. based on observation and participation in a given festival, interviews with the organisers, participants and spectators, or an analysis of archives, must take place in the case of a festival, which is the main object of study. Also, research conducted by students (e.g. to get credit at the end of term, or for BA or MA thesis) must include field study. As a result, the students gain more practical experience. Besides, they develop the skill of deeper cognitive and effective learning. Thus, they are motivated to understand the meaning of the learning content by connecting ideas with specific examples. This might be contrasted with a superficial approach to learning, which involves memorising theoretical information (Remmen and Frøyland 2014; see also Marvell et al. 2013).

After obtaining research material in a field study, it must be processed in a "laboratory" way. It is necessary to create databases, sort out the results by means of statistical analysis, describe them, draw maps (cartographic methods), etc. This type of study is called laboratory analysis. In some situations, it also includes obtaining data from offices or various institutions, as well as literature analysis, i.e. a library query (Cudny et al. 2011).

Examples

This subchapter will present individual research methods used in festival studies. They come from different, mostly social sciences. Festivals as social, economic and temporal spatial phenomena usually undergo an anthropological, sociological, economic and geographical analysis, and the research methods described below come from these scientific disciplines. They are also commonly used in event studies, which are a multidisciplinary science and as such are mostly based on methods borrowed from other sciences.

Before analysing any scientific problem, it is necessary to review relevant literature. An overview of the literature and Internet sources should precede all other studies and the most important conceptual work on the research project, including the choice of research methods. This comes from the fact that first, we must become familiar with the current state of the art concerning the issue in question. This will enable us to learn the definition of a given phenomenon, its history and present typology, the research approaches and the most frequently applied research methods. A literature review will prevent us from "reinventing the wheel", i.e. deliberating over problems the solutions to which have already been presented in other publications. Moreover, a literature overview will enable us to see the results of studies concerning the given problem as well as compare our study results to other analyses. Based on larger data, it is possible to make broader comparisons and generalisations, leading to more general conclusions, in the form of theories, typologies, models of phenomena, etc. This is the way of conducting an analysis which will stimulate the development of science and facilitate making broader generalisations (Fig. 3.1). Due to the role of literature analysis in the proper planning of research work, choosing research methods, formulating comparisons and finally drawing wide-ranging conclusions, it should be regarded as one of the basic laboratory research methods, also in the analysis of festivals (Table 3.2).

Fig. 3.1 The procedure of scientific cognition. *Source* Author's compilation

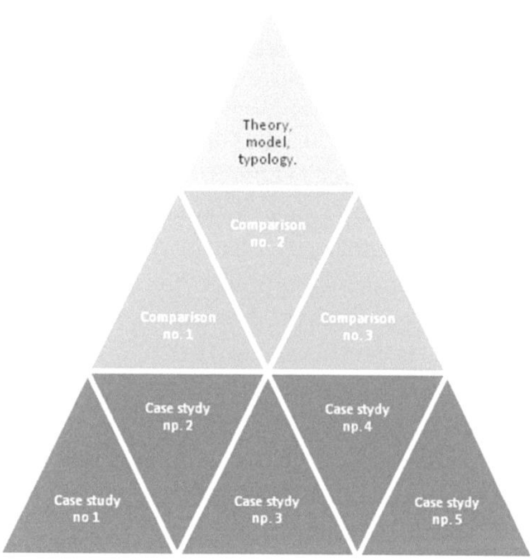

Table 3.2 Selected festival studies methods

Analysed problem	Research method
Scientific literature	Factual preparation to conducting research. Creating research programme, defining terminology, choosing research methods
Organisational structure of festivals, financing, employment	Analysis of promotional materials and Internet websites concerning the festival
	Query and interviews with the organisers and institutions supporting festivals (e.g. the city council, sponsors), festival workers and volunteers
The course of festival, realising the programme	Participant observation
Segmentation and motivations of festival visitors	Questionnaire or semi structured interview conducted among festival visitors
Perception of events by the inhabitants of host cities and regions	Questionnaire or semi structured interview conducted among the inhabitants of the host area
Economic impact of festivals on host areas	Survey including the issues of the visitors' expenses, research based on surveys conducted in firms and analyses of tourist traffic based on measurements, e.g. noting down registration plate numbers. Analysis of the advertising value equivalent, resulting, for example, from the presentation of the city in the media while discussing the festival organised there
The influence of festivals on the natural environment	Measurements and structured interview analysing the influence of festivals on the condition of air, water, noise intensity, the amount of waste, ecosystem balance, etc.
Analyses of festival distribution in space and time	Maps of the spatial distribution of festivals, maps and charts showing the changes in the distribution of events in time
Processing the collected information	Statistical analysis, descriptive models, monographic methods, SWOT and BCG methods in a study of a festival or a group of festivals

Source Author's compilation

The next popular laboratory method used in festival analysis is the study based on Internet sources (Table 3.2). Currently, a lot of information concerning, e.g. the organisers, history or programme of various types of events, can be found on the Internet. The majority of festivals have their websites. Moreover, many of them have professionally administered accounts on Facebook or Twitter (e.g. The Film Festival in Cannes or Venice). The accounts include information, festival programme, promotional materials and interviews with the participants. Websites or portals of this kind are a tool for communication between festival organisers and its spectators, fans, as well as the media and business (e.g. film) representatives. For the researchers concerned with festivals, the Internet is an invaluable source of information about things such as the structure, history or programs of events.

Besides, Internet websites may also be used for conducting surveys, e.g. among the fans or participants of a given festival. A questionnaire or a link to a website where it could be filled out can be placed, for instance, on the festival profile on Facebook; it may be sent to the festival fans registered on its mailing list, etc. The possibilities of using computer technology to collect information regarding festivals and reach their participants and organisers are practically limitless.

The next method is the query, which involves acquiring material from archival sources at offices and institutions (see Cudny et al. 2011). In the case of festivals, the query is a method used for studying the organisational structure, employment and financing of events (Table 3.2). We must reach the festival organisers' office and obtain the information we need, based on documents, such as reports, balances or contracts stored by the organisers. We may also obtain various types of documents at the council office of the city which supports a given festival financially and organisationally. In this case, a festival which is regularly supported has the obligation (written in the promotion agreement) to submit reports, which contain information about the organisers, budget, number of festival visitors, its programme, history, etc. Sometimes, these documents may also be obtained by post or e-mail (copies, scans). Then, the query may be regarded as a laboratory method.

Apart from the information obtained from the documents provided by the organisers or the supporting institutions, the data regarding the organisational structure, financing or employment structure of the festival can be acquired during in-depth interviews. Interviews are conversations with competent representatives of the organiser. With this method, it is worth paying attention to some practical advice. First, all conversations should be started from the top. The first person to talk to is the festival director or manager, as in this way we will get permission to conduct the study. Besides, the subordinates will then be instructed by their boss to provide us with any materials, information and assistance we need for the study. Such an approach will make it easier for us to proceed; moreover, if we do it this way, we will not be brushed off by incompetent or uncooperative people, who work on lower positions.

It is also important to be well prepared to all conversations. The theme of our work should be well thought out, and a list of questions to ask during our conversation should be carefully compiled. We should have a notebook, or (with the interlocutor's permission) the interview may be recorded. We should come to the meeting in person; any additional questions can be asked over the phone or e-mail. A proper, professional approach to the study in the form of an interview with the festival boss and workers makes us favourably perceived by the organisers, which will make it easier for us to cooperate and obtain the information needed for the study.

While studying a festival, it is necessary to monitor the course of the event, the realisation of its programme and the behaviour of the guests, spectators and organisers (Table 3.2). To do this, we can use the observation method. It is an extremely important research method used for verifying the number of participants, checking the standard of organisation during the event and watching the reaction of the invited guests and spectators. It is the basis for forming a general opinion about the atmosphere of the event and about its practical functioning. There are two forms of observation: indirect and direct. The first method does not require the researcher to take part in the process he/she studies. It is based on descriptions, information, reports coming from observing other people or institutions (see Cudny et al. 2011).

The method which is most useful or even indispensable in festival studies is direct or participant observation. It is a method popularly used in ethnography, based on the assumption that the real cognition of the lives of other people is possible only "from the inside". It involves the researcher becoming a part of a given community by taking part in its everyday life and by building relations with people, as well as watching and recording impressions in the form of notes, photographs, videos, voice recordings, etc. Graphic materials (e.g. photographs) obtained during observation may become an important illustration of the text. The researcher first has to gain access to a particular community, then lives/works among the people he/she studies and finally travels back to the research centre in order to analyse the research material (Flowerdew and Martin 2013). Another definition presented by Cudny et al. (2011, p. 30) says that "scientific observation is an accurate and directed registration of observations concerning a given object, phenomenon or process, taking place in natural conditions, without any intervention on the part of the observer (Leszek 1977). As pointed out by Urban and Ładoński (2001), in order to conduct proper observation, the following must be established:

1. The aim of observation;
2. The kind, location and character of the observed phenomena or objects;
3. The way of conducting observation and the rules of recording its results;
4. The place, duration and order of the observation;
5. The required number of observations;
6. The character of potential factors disturbing observation;
7. Ways of controlling and interpreting the results of the observation".

A major research theme in the festival analysis is the audience (see Sect. 3.1). In this case, the analysis regards the structure of the audience at a given festival as well as their motivations to attend it (Table 3.2). This type of analysis is conducted first of all in the event studies. Research on the festival audience structure is referred to as festival audience segmentation analysis (see Formica and Uysall 1995; Grant and Paliwoda 1998; Nicholson and Pearce 2000). The point is to define who the spectators are, i.e. to establish their demographic and social structure, so we analyse the spectators' sex, age, education and vocational structure, as well as their place of permanent residence (see Cudny and Stanik 2013; Cudny and Ogórek 2014). This kind of research enables us to define who they are, what social and vocational group they represent and where they come from.

Why are these research questions significant? They are important from the point of view of the event organisers, sponsors and other institutions supporting the festival. It is essential for them to learn about the audience structure, because it allows them to develop a better promotional strategy for the festival and to adjust its programme to the spectators' needs. It can be done if they know the age and education structure of the visitors; at the same time, it is possible to assume what environments they come from and what media they make use of. Such an analysis also checks to what extent the festival is a tourist asset and how much it is directed to the local community. This issue is studied mainly on the basis of the question concerning the spectators' place of residence. In this case, the questionnaire may include additional questions, e.g. whether they have come specially for the festival or are attending it by the way, by coincidence, whether they have come alone or with family or friends and how long they are going to stay. Such additional questions allow us to assess how important tourism generated by the event is for the local community. Moreover, we also obtain some information regarding the vocational structure of the spectators. This allows us to see to what extent the event is attended by professionals involved in a given type of activity (e.g. art).

The next issue in festival studies concerns the spectators' motivations to attend the events (Table 3.2). Studying motivations is important from the practical point of view; it provides us with the information about what persuades people to take part in them. As a result, we may check whether the festival offer is adequate or whether it should be better adjusted to the spectators' motivations. It is particularly important for the organisers. This analysis becomes even more valuable if we combine it with an analysis of the level of satisfaction with the festival. In this case, the questionnaire includes additional evaluative questions, e.g. whether the spectators liked the event, what should be changed in the future (e.g. the venue, information, promotion and program) or whether the spectators are planning to attend the festival in the future.

According to Cudny and Stanik (2013, p. 83), "Human behaviours are stimulated by different, more or less rational motivations. Festival visitors are also guided by certain motivations to attend an event". According to the encyclopaedic definition, motivation is "a process of psychological regulation, giving a behaviour energy and channelling it towards achieving an aim; also a permanent tendency (pursuit) to do activities oriented towards achieving a given aim" (http://www.pwn.pl). There are many publications, mainly connected with event studies, which concern the spectators' motivations to attend festivals. These works present the motivations of both the spectators who are tourists and those inhabiting the host area. As shown by Cudny and Stanik (2013), there are works devoted to the analysis of spectator's motivations to attend art (Schneider and Backman 1996; Nicholson and Pearce 2001; Van Zyl and Botha 2004; Lee et al. 2004; Bowen and Daniels 2005), religious (McDowall 2010), or food and wine festivals (Yuan et al. 2004). Based on the review of the literature concerning motivations to attend festivals, Cudny and Stanik (2013) as well as Cudny and Ogórek (2014) distinguish six main groups of motivations, which most often persuade spectators to attend festivals (Table 3.3).

The group of motivations related to culture and art is typical of people visiting art festivals (e.g. film, theatre, music and ballet). Modern developed societies value intangible culture; for many of them, contact with culture is an important element of life, even on daily basis. Many people are fans of specific arts, individual artists or bands. Therefore, the number of persons who want to take advantage of cultural events such as art festivals is growing. What is more, these festivals are strongly promoted, for example, by the city and regional authorities, supported by sponsors

Table 3.3 Selected groups of motivations to attend festival events

Group of motivations to attend a festival event	Selected motivations
Related to culture and art	To experience culture and art
	To extend knowledge about art
	To become familiar with a new cultural event
Related to the event attractiveness	To enjoy the festival atmosphere
	To see new, interesting things
	To take part in festival life
Related to originality, novelty	To gain new experience
	Out of curiosity
	Looking for benefits and experience enriching myself
Related to the escape from the everyday routine	To escape from the tedious routine of every day
	To unwind
	To slow down the pace of everyday life
Related to social contacts	To meet new, interesting people
	To meet people with similar interests
	To meet people from my occupation group
	To go out with friends
	To meet interesting people from abroad
Related to family life	To spend more free time together with the family
	To show the family an interesting place
	To introduce interesting people to the family

Source Author's compilation based on Cudny and Stanik (2013), Cudny and Ogórek (2014)

and presented in the media. Professional promotional campaigns are prepared by firms concerned with the organisation and management of commercial events, e.g. music festivals. People who have a lot of free time and financial means look for interesting experiences related to art and that is why they attend festivals.

The second and the third group of motivations (Table 3.3) include the willingness to take part in an attractive, interesting event which has the aura of novelty and extraordinariness. Once again, this can be related to the theory of experience societies described earlier (see Sect. 2.2). Searching for new, unusual, exciting events and experiences is the main motivation here. Festivals are something exceptional (they are often held only once a year); they often present foreign artists who are normally inaccessible to the spectators. Moreover, they have a unique atmosphere and are interesting, special events, full of new people (artists and spectators), presenting and promoting new forms of art, new ideas, discoveries, etc. Festivals are attended by famous artists, journalists and celebrities. Events related to exploration and travel present the latest discoveries, extreme sports, expeditions and the people taking part in them. All this is a reason why such an event has a unique atmosphere, as well as attracts people with its interesting and unusual contents. People visit festivals for cognitive purposes as well, in order to increase their knowledge about a given art, listen to fascinating stories about the world told by travellers or learn about scientific and technological novelties (science and technology festivals).

At the same time, the events discussed here are a perfect opportunity to escape from the everyday routine. The pace of contemporary life is fast and exhausting; people want to get away from the dreary everydayness, and this is why they look for various attractions. One of the theories developed in tourism is the push and pull factor theory. The push factors are responsible for pushing people out of their regular surroundings and they include tiredness, boredom with everyday routine, etc. On the other hand, the pull factors, which attract people to a tourist destination, include the possibility to rest and relax, peace and quiet, and a new attractive environment. (see: Iso-Ahola 1982). Apart from tourism (e.g. festival tourism), these factors are active also in our everyday lives. We want to rest, get away from it all, and spend our free time in an interesting way, e.g. together with the family or friends. To do that, we search for additional attractions, and festivals can provide us with attractions of this type in the evenings or at weekends in our everyday environment.

Festivals are places where people make and develop social contacts. They integrate acquaintances and persons who share common interests, e.g. related to the theme of the festival. For instance, the place promoting contacts among persons interested in comic books and computer games is the International Festival of Comic Books and Games—the largest in Poland and organised annually in Łódź (see http://komiksfestival.com/en/). Finally, festivals are places of contact for the world of business. For example, the Cannes Film Festival is visited not only by film fans from all over the world, but also by famous actors, film directors, cinematographers and other film-makers. It is also attended by journalists writing film reviews for renowned international journals, as well as by film producers and agents. It is where business contacts are made or even contracts are signed for the purchase and distribution of individual films. Therefore, it can be said that festivals are places of social capital development (see Arcodia and Whitford 2006) and at the same time, they are obligatory points of passage (see Sect. 2.1).

The next group of motivations concerns family life. They are in a way similar to those from the previous group, as they are related to social life. In this case, the point is to spend more time with family, e.g. children, the spouse and parents. Everyday life, often overloaded with duties and chores, sometimes makes contact with our closest family members difficult. Therefore, going to a festival together is an opportunity to spend time with them in an interesting environment.

Research concerning the segmentation and motivations of festival audience is usually conducted with the use of sociological methods, such as questionnaire survey or semi-structured interviews. The survey involves creating a questionnaire containing questions regarding, e.g. the demographic–social features and motivations of festival audience. Next, these questionnaires must be delivered to respondents, and after they fill them out—collected. The researcher has no possibility to explain the questions or answer the questions asked by the respondents, or ask them additional questions. The respondents receive a written list of questions in the form of a questionnaire. It must be remembered that the questionnaire is usually filled out without the researcher's participation, so it may have a number of disadvantages. For instance, it is not always completed by the person for whom it was intended. Besides, different people fill it out in different conditions, e.g. at different places and times. As it was mentioned above, the questionnaire does not give the researcher the opportunity to have a discussion with the respondent and makes it impossible to observe him/her during the examination. Another difficulty while compiling questionnaires is formulating the required number of forms, which is hardest to achieve in self-returnable questionnaires, i.e. those which are returned by the respondents at designated places or sent back by mail. In such cases, we should plan to distribute more forms than the number of questionnaires we require. The research experience shows that in the case of questionnaires filled in personally by the respondents, receiving 20 completed forms out of 100 that have been sent out is quite a good result. Depending on the way of delivering the questionnaire to the respondents, we may identify the following types of surveys: mail, press, a questionnaire attached to purchased products, radio survey (the text is transmitted through the radio and the answers are sent by mail), TV survey, a hand-distributed questionnaire (the researcher distributes and collects the questionnaires personally), telephone survey and a questionnaire distributed through the Internet, e.g. by e-mail or placed on a web page where the users may fill it in online or print it and send it back (Cudny et al. 2011; Veal 2006).

Semi-structured interview is characterised by its length, depth and structure. This method takes more time than the questionnaire survey. The researcher may ask more detailed questions, which can be more complex—not only closed, but also open or mixed. It is possible to discuss the questions and exchange questions with the respondents, and it is possible to observe them during the interview. Semi-structured interview is therefore less structured than the questionnaire survey. However, this does not mean that we do not use a formalised questionnaire form in this case. On the contrary, we may use a questionnaire similar to that used in the survey described above, though the questions are more complex and the way they are asked is different. During an semi-structured interview, the questioner asks the questions and writes down the answers, possibly discussing some of them further. It is also possible to run the study without a detailed questionnaire, which can be replaced with a list of problems to talk about. This method is called in-depth interview, however, is used with a smaller number of respondents, e.g. when talking to the festival staff, volunteers or organisers. For the analysis of segmentation or motivations, we should choose a unified interview questionnaire (Veal 2006; Cudny et al. 2011).

When studying festival visitors, it is best to use the method which involves distributing a questionnaire about a given festival by hand among its visitors. First, it is necessary to choose a suitable place for distributing the forms so that it is possible to quickly reach the maximum number of visitors. It might be the festival reception area, where they come to buy tickets or festival programme brochures. A good practice is handing in questionnaires together with the programmes or the tickets. While handing in a questionnaire, the researcher should introduce themselves and ask whether a given person has come to the festival and whether they will agree to take part in the survey. If the answer is affirmative, we hand in the questionnaire, briefly describing it and explaining where it can be returned (e.g. at the festival reception desk). It is not possible to distribute the questionnaires among all the spectators, especially at large festivals attended by several thousand or more people. We try to approach, for instance, every fifth person, differentiating among the respondents and choosing people of different sex or age, if possible. This kind of selection is called random sampling and involves choosing a specific number of respondents or a specific percentage of the whole group, in a totally random (incidental) way. Such research should be treated as random study. This means that the respondents are haphazardly selected people, available for study at a given moment, who do not reflect the structure of the festival audience.

Another type of sampling, which guarantees a higher representativeness of results, is the systematic random sampling. The group of respondents (smaller than the whole collective), called the sample, must contain the same percentage of individual groups of respondents as the statistical population. Therefore, first, we must know the percentages regarding the sex, age and education structure. In our sample, the structure of respondents must correspond to the structure of the statistical population, e.g. of all the visitors to a given edition of the event. The number of the questionnaires is not the most important thing, though in some works it is assumed that it should make up at least 1 % of the number of the statistical population (with large collectives). In the study of festival audiences, this condition usually cannot be fulfilled. It comes from the simple fact that the organisers do not have such detailed data concerning spectators. Moreover, even if they have it, they cannot provide it until after the festival (for sampling procedures see Janeczko et al. 2002).

The questionnaire survey and semi-structured interview is also the basic research method used in the analysis of the perception of events by the inhabitants of host cities and regions. Every event exerts significant (positive and negative) impacts on the local community, e.g. it generates jobs, modifies prices, increases income from trade and may cause traffic problems or riots. On the other hand, the people inhabiting the host area are its social base. They may support the event organisationally (e.g. as volunteers) and approve of it or not by putting pressure on the local authorities, and finally, they may even protest against the event which they do not accept. Considering these strong interactions, it is important to study the inhabitants' opinions regarding a given festival event.

Examples of such studies can be found, among others, in the Australian geographical literature. One of the works presents the town of Tamworth, which has hosted the Country and Western Music Festival since the 1970s. The event attracts about 60 000 visitors every year and recently has become the main element of the development strategy for this destination. The whole promotional campaign of Tamworth is based mainly on the connotations with the countryside, country music, masculinity and Anglo-Saxon traditions. An interesting issue is the perception of the festival by the local inhabitants. In the beginning, a part of the inhabitants did not accept the festival; some even declared leaving the city for the duration of the event. With time, however, negative reactions weakened and the event was gradually accepted. The inhabitants started to perceive the festival as a promotion factor, bringing profits and generating economic development. However, they still could see dangers, such as the inconveniences in everyday life resulting from a large number of tourists flowing into this small town, strengthening the image of the town as a "redneck" or "hip" destination, associated with the countryside style, etc. (see Gibson and Davidson 2004). Some evidence of business support provided by the host community for another music festival, dedicated to Elvis Presley, is indicated in another Australian publication by Brennah-Horley et al. (2007).

An interesting example of the analysis of the perception of festivals organised in Łódź—a large post-socialist Polish city—can be found in a publication by Cudny et al. (2012). Based on 1200 questionnaires, the authors investigated the inhabitants' perception of the festivals held in Łódź, how they evaluated the economic and promotional impacts, as well as the organisation of the events, the support given to festivals by city authorities, etc. The respondents were also asked which festivals they regarded as the most important for the city and which ones they had attended in the previous years. The basic research method was the semi-structured interview. Before the main investigation, the authors conducted a test study. It included a group of 50 randomly chosen respondents, and its aim was to check whether the questionnaire had been constructed properly. The main investigation was conducted among adult inhabitants of Łódź. The interviews were realised in the city centre, in public places. The interviews were conducted on six different working days. Every fifth passer-by was approached and asked whether they wanted to take part in the study. During the survey, the questioners tried to ask a similar number of men and women, as well as people of different ages. The results of the study let the researcher evaluate the attitude of city inhabitants to festivals (generally positive), as well as investigate the opinions concerning their role in the development of the city. They also showed how the inhabitants perceived the level of cooperation between the local authorities and festival organisers. Furthermore, based on the study results, it was possible to find out which of the ca. 60 festivals organised annually in Łódź are most popular among the inhabitants.

The economic influence of festivals on the host areas and host communities is the next crucial element of festival studies (see Table 3.2). It is highly significant for the organisers as it defines the profitability of organising an event from the economic point of view. It affects the decisions taken by the institutions supporting festivals (local governments, sponsors), giving them an idea of the real economic consequences. Finally, it is interesting for host communities because it lets them discover the actual economic influence of events in terms of the income of local firms or creation of new jobs. All these groups are very important because they make key decisions concerning the support for a given event (Jago and Dwyer 2006). There are many methods which can be used to research this aspect (Table 3.2). The economic impacts of festivals and other events can be divided into two categories. The first one includes direct effects, and the other one—the indirect ones. Analysing the first group is quite easy, while calculating the indirect effects is much more difficult. It requires the use of econometric methods based on estimated calculations. Calculating the direct impact involves defining the amount of income gained from organising a festival at a given host destination. It may be studied with two methods. The first one is based on the research conducted among festival organisers (business sector surveys), and the other one makes use of the analysis of the event participants' expenditures (consumer/participant survey). In the first case, we must contact the event organisers (or firms working for the organisers) in order to establish the balance of costs and profits drawn from the organisation of festivals. The costs include the money spent on artists (e.g. in the case of art festivals), the lighting, renting the venue, employing the staff, hiring security, the cost of cleaning and waste disposal services. The profits include income from selling tickets (if the event is ticketed) and fees for broadcasting the festival in the media, for presenting commercials during the festival (e.g. on LED displays). It is also possible to conduct research among firms offering services to festival tourists but not working directly for the event, such as hotels, restaurants or taxi companies. Services provided by them are used by festival tourists. Therefore, it is possible to turn to them with the question whether they noted an increase in their income during the festival. In this case, however, we must take into account other, for example, seasonal changes in the tourist traffic. For instance, if the festival is organised during the holiday season, it is difficult to tell the difference between higher profits caused by the event and the ordinary increase in turnover at that time. If the analysis is based on the information provided by the organisers, we should remember that obtaining it may sometimes be very difficult. It comes from the fact that many commercial companies organising events do not want to reveal financial data to the competition. Moreover, they are often afraid that they will not receive further support from the sponsors and advertisers when it turns out that the event has attracted a smaller number of spectators than planned. Another problem may be the fact that the data obtained from the event organisers are secondary data, which does not always meet the researcher's needs (Janeczko et al. 2002).

The other method of assessing the economic impacts of festivals is the customer/participant survey. It is a direct method, as we personally reach the persons who generate profits and we obtain "first hand", unprocessed data. We may ask about the details which are interesting to us, e.g. whether a given visitor has arrived specially for the event or not, whether he/she has arrived alone or with family (or friends) and how many persons accompany him/her. All these allow us to define the economic impacts of the festival more precisely. It is easier to ask participants about their spending and behaviour patterns than to ask business people for their sales figures. The questionnaire may contain questions regarding the length of stay and the amount spent on goods and services (tickets, accommodation, food and drinks, petrol, etc.). The study of festival visitors, however, is not without a fault. Firstly, we must be prepared for the respondents' resentment towards participation in the study; they may not like the idea or have time to fill in the questionnaire or take part in an interview. Secondly, studies like that are time-consuming and costly. The most popular method in this kind of studies is a questionnaire survey or a semi-structured interview. They will not be described in detail in this part of the chapter, as they have been characterised before (for details see also: Janeczko et al. 2002; Madden et al. 2002). We may approach this type of study in different ways. First, we may conduct it at various times, i.e. before, during or after the event. Before the event, we may investigate the potential willingness to visit the festival, e.g. among the inhabitants of the city or region. During the event, it is easiest to choose the respondents, because we may turn directly to the persons arriving at the festival venue. After the event, we may turn to the persons who have logged into the electronic system of guest registration and ticket sale. In this case, depending on the data obtained from the organiser (address, e-mail, phone number), we may send them the questionnaire or call them. Moreover, we may use various survey techniques, such as the distributed questionnaire, phone survey, questionnaires sent by mail or e-mail and semi-structured interviews (Janeczko et al. 2002).

It is important that the evaluation of the economic influence of an event should be comprehensive. To achieve this, we can combine a number of methods, e.g. studies conducted among a group of organisers with those carried out among visitors. What is more, we should analyse both the direct and indirect economic impacts. If we want to define the indirect impacts, we use a method checking how the money earned during the festival "is distributed across the economy" (see Janeczko et al. 2002, p. 27; Jago and Dwyer 2006, p. 25). As for the expenditures on the organisation of the event, we must also consider the sums spent on building festival facilities, e.g. performance and festival halls, or developing the infrastructure, e.g. transport infrastructure, for festival purposes. We should also take into account whether this infrastructure has been used only once or perhaps it will be used many times during a regular event or other similar events, or whether it will be used by the inhabitants in their everyday lives. We should check the costs connected with the influence of festivals on the natural environment, e.g. an increased water intake, bigger pollution, waste production and the resulting higher costs of storing and utilisation. A similar price paid by local communities for organising events is the increase in the prices of basic goods, e.g. food, usually noticeable not only during the event, but also after it ends (for details see Jago and Dwyer 2006). Apart from that, we should examine the advertising value equivalent and the role of festivals in shaping the brand of a given destination or firms sponsoring the event. The advertising value equivalent is the due cost of the publication or broadcast of a given message (e.g. concerning a given host city) as if it were a paid advertisement. Sponsors and city or regional authorities often decide to support, co-finance unprofitable festivals because they are often presented in the media. Such sponsoring may be profitable because if firms or destinations wanted to buy comparable airtime on television or the radio, or a press or Internet advertisement, they would have to pay larger sums. All these elements should be taken into account in order to evaluate the real economic balance of the organisation of a given festival (for details about the advertising value equivalent see Macfarlane and Jago 2009).

Festival events have a significant influence on the natural environment, which has been presented in several publications (see Jones et al. 2008; Kulshrestha et al. 2004; Cudny 2013). The influence on the natural environment is most frequently connected with the additional burden caused by intensive tourist traffic. It particularly concerns large events, attracting large numbers of tourists. Environmental consequences of mass events (including festivals) and the methods of studying them have been thoroughly discussed in Australian studies (see Jones et al. 2008).

Considering the significance of these phenomena, it is very important to evaluate the influence of individual events on the environment (Table 3.2).

It should be done in every individual case, so that the event programme and its services could burden the environment to the least possible extent. An analysis of this kind may be based on the researcher's own studies (e.g. noise level measurement) or (more often) on a properly constructed in-depth or semi-structured interview. It should be addressed to different stakeholders (e.g. organisers, host organisations, such as the city council, affiliated organisations) responsible for the organisation, service and management of the event. The interview should include questions concerning the following elements (Jones et al. 2008):

- The number of vehicles driving through a given locality on the festival days. The organisations which should be turned to are the police, traffic management office and firms managing parking spaces.
- The consumption of energy and water during the event. These data can be obtained from the firms providing service at a given event; the data may be delivered in the form of electricity bills; it may include information about the amount and distribution of power and the consumption of fuel for power generators working for the festival. Perhaps, the data could be obtained from the utility companies providing energy for the host area as long as the festival uses network electricity and not generators.
- The amount of waste produced during the festival. In this case, it is worth turning to firms concerned with city cleaning. We may also approach the organisers or the cleaning firms hired by them. The reports on waste should include the issues of waste sorting and recycling.
- The toilet facilities used during the festival. Visitors may use public toilets, but because there are not many of them, temporary (movable) toilets are used more often. They are provided by companies specialising in toilet rental and waste utilisation. The number of toilets and the amount of sewage, as well as the rules of utilisation, should be included in the analysis of the environmental impacts.
- The noise. In this case, we should make measurements during the festival; it can be done by the researchers, the representatives of the organisers or supporting institutions.

Apart from collecting data, it is also important to compare the level of pollution during the festival with that at the time when the event is not taking place. The obtained results may also be compared to the pollution norms for a given area. The study results may be next presented in the form of a summary evaluating the influence of a given festival or a group of festivals, or other events on the natural environment. On the basis of the information acquired during interviews, the researcher ascribes the level of influence on the environment (e.g. high, medium, low) to a given feature (e.g. the amount of precipitations and noise). This method is particularly useful in the analysis of a group of events whose environmental influence may be then compared, e.g. in a table (Jones et al. 2008).

After collecting data by means of different kinds of field study methods, such as observation, query, surveys or interviews, it should be processed and presented in the form of laboratory studies. These methods include cartographic, statistical, descriptive and monographic methods, as well as SWOT or BCG analyses. For presenting the distribution of festivals in space and time (Table 3.2), cartographic and graphic methods are used. Cartographic methods involve making all kinds of maps, e.g. presenting the spatial distribution of festivals. Examples of such graphic presentations in the form of maps can be found in publications by Visser (2005) or Gibson et al. (2010). Thanks to the signature or dot technique, maps may present the distribution of festivals, e.g. over a given region of the country, or the distribution of facilities used for festival purposes within the city space, e.g. in the form of a cartogram (see Adamus and Pluch 2012).

The signature method involves presenting elements on the map in the form of symbols, called signatures. Their location on the map marks the occurrence of a given phenomenon at a given place, and the shape of the signature says what phenomenon it is (see Cudny et al. 2011). This method is used mainly with qualitative (immeasurable) phenomena, which include festival location. Festivals can be divided into different types, e.g. according to their themes. Using different signatures, it is possible to present the change of location and the number of festivals in different years.

The dot method shows the distribution and number of, for example, festivals. A dot is placed at the point of festival location, and its size may signify their number (from one to many). Moreover, the number of festivals may be presented with the use of cartographic quantitative methods, e.g. on thematic maps or cartograms presenting, e.g. the number of festivals within one area unit (e.g. in a *gmina*/commune or region). The distribution of festivals in time (and in space) may also be presented by means of graphic elements, such as charts and tables.

Interesting laboratory methods developed in economic sciences and management, which can be used in the analysis of festivals are methods known as SWOT and BCG. Let us first look at the SWOT analysis, often called an analysis of opportunities and threats. It is often used for the analysis of enterprises in order to define external elements functioning around the enterprise, i.e. the opportunities which may cause its development and the threats which constrain the development. Internal factors, i.e. weaknesses and strengths are also analysed. This enables the researcher to define the elements triggering the growth of the enterprise as well as those internal and external factors which hamper it (for details about SWOT analysis see Pahl and Richter 2007). A similar analysis can be carried out with reference to a festival, whose organisation may be treated as a particular kind of a business enterprise. Also here, we deal with positive and negative internal and external factors. Studying them may help to improve the functioning and effectiveness of the event (Table 3.4).

Another interesting method deriving from management sciences or the management techniques portfolio, to be exact, is the growth–share matrix proposed by the Boston Consulting Group (BCG). It may be used for the evaluation and management of groups of products within the framework of a given organisation, e.g. an

Table 3.4 SWOT analysis of Łódź Fashion Week held in Łódź (Poland)

Strengths	• Efficient organisers (private persons) • Help and support from city authorities and private sponsors • Interesting offer of shows, workshops and exhibitions • Good offer for young, promising designers, on the other hand an interesting event for well-established fashion houses and clothing companies (hunting for new trends and talents) • Presence of distinguished guests, famous fashion designers, journalists	• Dependence on rented premises, lack of own premises • Dependence on external financing	Weaknesses
Opportunities	• Belongings to the international system of Fashion Week events • International cooperation wit fashion institutions • Location at a large textile and clothing industry centre of long tradition (Lodz) • Development of business tourism • Lack of competition from other such events in Poland • Large number of visitors from the region, the country and abroad • Interest of the media (also trade-related)	• Competition from other events of this type in the world (Fashion Weeks) • Global crisis causing smaller interest in fashion and limited funds from trade-related firms • Risk of losing sponsors and support of city authorities due to crisis (lack of self-financing) • Competition from other fairs and festivals in Lodz	Threats

Source Cudny and Rouba (2012, p. 55)

enterprise. It is about the assessment of these products based on two elements: the growth rate of a particular market and the organisation's share in that market (see Enz 2010; Griffin 2012).

Festivals held at a specific destination may be in this case treated as a group of products, which can be divided into various groups, using the principles of the BCG matrix. This type of analysis was for example carried out by Cudny and Rouba (2011) in the study of theatre and multicultural festivals (understood here as events encompassing various domains of non-material culture) held in Łódź. The authors divided festivals into several basic types on the basis of their most important characteristics, such as market share, the expenditure necessary to produce and promote them and the number of spectators—participants. Following the rules of the BCG matrix, they distinguished the following types of products: cash cows, stars, question marks and dogs. "The cash cows have a major share in the market

Table 3.5 Growth–share matrix for Łódź theatre and multicultural festivals

	Large share in the tourist market	Small share in the tourist market
High market growth rate	Stars: • Spotkania Teatrów Miast Partnerskich (T) • Międzynarodowy Festiwal Sztuki Ulicznej „Trotuart" (T)*	Question marks: • Nowa Klasyka Europy (T) • Letni Festiwal Sztuki (M)** • Festiwal Sztuk „Cygańskie Impresje" (M)
Low market growth rate	Cash cows: • Międzynarodowy festiwal Solistów Lalkarzy (T) • Łódzkie Spotkania Baletowe (T) • Ogólnopolski Festiwal Sztuk Przyjemnych i Nieprzyjemnych (T) • Międzynarodowe Biennale Spotkania Teatralne „Terapia i Teatr" (T) • Festiwal Szkół Teatralnych (T) • Międzynarodowy Festiwal Komiksu i Gier (M) • Międzynarodowy Festiwal Fotografii (M) • Łódź Biennale (M) • Międzynarodowe Triennale Tkaniny (M) • Międzynarodowe Triennale Małe Formy Grafiki (M)	Dogs: • Łódzkie Spotkania Teatralne (T) • Ogólnopolski Przegląd teatrów Dziecięcych (T) • ŁóPTA Łódzki Przegląd Teatrów Amatorskich (T) • Międzynarodowe Warsztaty Folklorystyczne (M) • Kongres Iluzjonistów (M)

Legend
*(T)—Theatre festivals
**(M)—Multicultural festivals
Source Cudny and Rouba (2011, p. 15)

and require small means on promotion. The stars are new products, with a growing share in the market, still in need of considerable promotion. Question marks are just entering the tourist market. It is not known yet whether they will be effective or not, and they require intensive and costly promotion. They may transform into stars and then into cash cows, but they may also turn out to be misfits and disappear from the market. Dogs are tourist products which are slowly leaving the market, old-fashioned, but still fairly interesting and popular. When their further existence becomes unprofitable, they are withdrawn from the tourist market" (Cudny and Rouba 2011, pp. 54–55). A large group of urban festivals were evaluated in cited work in this way as regards their share in the market and further development opportunities (see Table 3.5).

An analysis of this type is important from the point of view of the culture development strategy in a given city. Authorities who usually support various festival events in their city may be interested in analyses of this type in order to evaluate their present impacts and further development opportunities. Such a study can be also useful for firms administering events, whose portfolio contains many events of varying share in the market. On the basis of the BCG analysis, we may determine which of them already bring profits, which require further support and which should be perhaps removed from the festival market.

Other methods of laboratory studies are statistical analyses, which involve calculating indexes, e.g. when analysing survey results. Calculating the structure of responses, percentages, the values of arithmetical means, medians, standard deviations, etc., is simple statistical methods which can be used for the analysis of festival-related phenomena. The next laboratory method indispensable in every scientific work is the descriptive method. It involves producing a text description of the studied phenomenon. One of its types is the monographic method, in which the description regards one chosen issue, e.g. an individual festival. However, this description must be thorough and comprehensive. Another laboratory method which can be used in festival studies is the model method. Generally speaking, it involves constructing simplified models of reality, which depict the key elements and features of a given process. It is a way to present issues connected with the functioning of a festival or its influence on the surroundings. The model method may be a good way to sum up the analysis of a festival or a group of festivals organised in a given area. In this part of the book, issues connected with the rules of constructing models and dividing them into different types will not be discussed in detail. A thorough description of this method can be found in Chap. 6, which includes the model of the roles of festivals in the management of contemporary urban space.

Study Tips

- Describe the main research themes in the analysis of festivals in different scientific disciplines.
- Explain what research methods are and why do we use them in science.
- Characterise how the research methods can be divided.
- List and describe the most important methods of studying festivals.

References

Abreu Novais M, Arcodia C (2013) Music festival motivators for attendance: developing an agenda for research. Int J Event Manag Res 8(1):34–47

Adamus J, Pluch M (2012) Wydarzenia kulturalne w mieście poprzemysłowym. Przykład Łodzi. Turystyka Kulturowa 9:35–54

Anderson K, Domosh M, Pile S, Thrift N (2003) A rough guide. In: Anderson K, Domosh M, Pile S, Thrift N (eds) Handbook of cultural geography. Sage Publications LTD., London, pp 1–36

Arcodia C, Whitford M (2006) Festival attendance and the development of social capital. J Convent Event Tour 8(2):1–18

Avraham E (2014) Hosting events as a tool for restoring destination image. Int J Event Manag Res 8(1):61–75

Axelsen M, Swan T (2010) Designing festival experiences to influence visitor perceptions: the case of wine and food festival. J Travel Res 49:436–450
Backman KF, Backman SJ, Uysal M, Sunshine KM (1995) Event tourism: an examination of motivations and activities. Festival Manag Event Tour 3(1):14–24
Bowen H, Daniels MJ (2005) Does the music matter? Motivations for attending a music festival. Event Manag 9:155–164
Bjeljac Ž, Lović S (2011) Demographic analysis of foreign visitors to the Exit Festival, Novi Sad. J Geogr Inst "Jovan Cvijić" SASA 61(2):97–108
Binns L (2005) Capitalising on culture: an evaluation of culture-led urban regeneration policy. Futures Academy, Dublin Institute of Technology, pp 1–8. http://arrow.dit.ie/cgi/viewcontent.cgi?article=1004&context=futuresacart&sei-redir=1&referer=http%3A%2F%2Fscholar.google.de%2Fscholar%3Fstart%3D10%26q%3Dculture%2Bled%2Bregeneration%2B%26hl%3Dpl%26as_sdt%3D0#search=%22culture%20led%20regeneration%22. Accessed 03 July 2012
Brennah-Horley C, Connell J, Gibson C (2007) The Parkes Elvis Revival Festival: Economic development and contested place identities in rural Australia. Geogr Res 45(1):71–84
Buch T, Milne S, Dickson G (2011) Multiple stakeholder perspectives on cultural events: Auckland's Pasifika festival. J Hosp Market Manag 20(3–4):311–328
Crang M (2002) Qualitative methods: the new orthodoxy? Prog Hum Geogr 26(5):647–655
Crang M (2003) Qualitative methods: touchy, feely, look-see? Prog Hum Geogr 27(4):494–504
Crang M (2005) Qualitative methods: there is nothing outside the text? Prog Hum Geogr 29 (2):225–233
Crompton J, McKay S (1994) Measuring the economic impact of festivals and events: some myths, misapplications and ethical dilemmas. Festival Manag Event Tour 2:33–43
Crompton J, McKay S (1997) Motives of visitors attending festival events. Ann Tour Res 24:425–439
Cudny W (2011) Film festivals in Łódź as a main component of urban cultural tourism. Bull Geogr Socio-econ Ser 15:131–141
Cudny W (2013) Festival tourism—the concept, key functions and dysfunctions in the context of tourism geography studies. Geografický časopis 6(2):105–118
Cudny W (2014a) Festivals as a subject for geographical research. Geografisk Tidsskrift-Danish J Geogr 114(2):132–142
Cudny W (2014b) The phenomenon of festivals—their origins, evolution and classifications. Anthropos 109(2):640–656
Cudny W, Gosik B, Piech M, Rouba R (2011) Praca dyplomowa z turystyki. Podręcznik akademicki, Łódzkie Towarzystwo Naukowe, Łódź
Cudny W, Korec P, Rouba R (2012) Resident's perception of festivals—a case study of Łódź. Sociológia – Slovak Sociol Rev 44(6):704–728
Cudny W, Rouba R (2011) Theatre and multicultural festivals in Lodz as a free time management factor among the inhabitants and tourists in the post-industrial city. Acta Geogr Univ Comenianae 55(1):3–22
Cudny W, Rouba R (2012) Lodz Fashion Week as an example of a business event. Acta Geogr Univ Comenianae 56(1):45–58
Cudny W, Stanik E (2013) "Man in danger" Media Festival in Lodz—the structure and motivations of the festival visitors. Ann UMCS Sectio B (Geographia, Geologia, Mineralogia et Petrographia) LXVIII(1):69–90
Cudny W, Ogórek P (2014) Segmentation and motivations of the attendees' of the Mediaschool Festival in Łódź, Poland. Bull Geogr Socio-econ Ser 24:41–56
Czepczyński M (2007) Podejścia badawcze w nowej geografii kultury. In: Maik W, Rembowska K, Suliborski A (eds) Geografia a przemiany współczesnego świata. Wydawnictwo Uczelniana WSG, Bydgoszcz, pp 203–211
David S (1995) A comparison of visitors' motivations to attend three urban festivals. Festival Manag Event Tour 3(3):121–128

Davies WKD (2015) Festive cities: multi-dimensional perspectives. In: Davies WKD (ed) Theme cities: solutions for urban problems. Springer, Chaam, pp 533–561

Delamare TA (2001) Development of a scale to measure resident attitudes toward the social impacts of community festivals, part II: verification of the scale. Event Manag 7:25–38

Delamare TA, Wankel LM, Hinch TD (2001) Development of a scale to measure resident attitudes toward the social impacts of community festivals, part I: item generation and purification of the measure. Event Manag 7:11–24

Della Bitta A, Loundon D, Booth G, Weeks R (1977) Estimating the economic impact of a short term tourist event. J Travel Res 16(2):10–15

De Bres K, Davis J (2001) Celebrating group and place identity: a case study of new regional festival. Tour Geogr 3(3):326–337

Derrett R (2003) Making sense of how festivals demonstrate a community's sense of place. Event Manag 8:49–58

Duvignaud J (1976) Festivals: a sociological approach. Cultures III(1):13–25

Duvignaud J (1989) The festive spirit. The UNESCO Courier XII:11–17

Eckman AK (1999) The revival of cultural celebrations in regional Sweden. Aspects of tradition and transition. Sociol Rural 39(3):280–293

Enz C (2010) Hospitality strategic management: concepts and cases. Wiley, Hoboken

Evans G (2005) Measure for measure: evaluating the evidence of culture's contribution to regeneration. Urban Stud 42(5/6):959–983. doi:10.1080/00420980500107102

Falassi A (1987) Festival: definition and morphology. In: Falassi A (ed) Time out of time. University of New Mexico Press, Albuquerque, NM, pp 1–10

Flowerdew M, Martin D (2013) Methods in human geography. A guide for students doing a research project. Routledge, Oxon

Formica S, Uysal M (1995) A market segmentation of festival visitors: Umbria jazz festival in Italy. Festival Manag Event Tour 3(4):175–182

Gamson J (1996) The organizational shaping of collective identity: the case of lesbian and gay film festivals in New York. Sociol Forum 11(2):231–261

Getz D, Frisby W (1988) Evaluating management effectiveness in community run festivals. J Travel Res 27(1):22–27

Getz D (1989) Special events defining the product. Tour Manag 10(2):125–137

Getz D (1991) Festivals, special events, and tourism. Van Nostrand Reinhold, New York

Getz D (2000) Developing a research agenda for the event management field. Events beyond 2000: setting the Agenda. In: Proceedings of conference on event evaluation, research and education, Sydney. July 2000, pp 10–21

Getz D (2010) The nature and scope of festival studies research. Int J Event Manag Res 5(1):1–47

Getz D (2012) Event studies. Theory, research and policy for planned events. Routledge, Oxon

Gibson C, Connell J (eds) (2011) Festival places: revitalising rural Australia. Channel View Publications, Bristol–Tnawanda

Gibson C, Connell J (2012) Music festivals and regional development in Australia. Ashgate, Farnham

Gibson C, Davidson D (2004) Tamworth, Australia's 'country music capital': place marketing, rurality, and resident reactions. J Rural Stud 20(4):387–404

Gibson C, Waitt G, Walmsley J, Connell J (2010) Cultural festivals and economic development in nonmetropolitan Australia. J Plann Educ Res 29(3):280–293

Goldblatt J (1990) Special events: the art and science of celebration. Wiley, New Jersey

Goldblatt J (2000) A future for event management: the analysis of major trends impacting the emerging profession. In: Allen J, Harris R, Jago LK, Veal AJ (eds) Events beyond 2000: setting the agenda proceedings of conference on event evaluation, research and education, Sydney July 2000. Australian Centre of Event Management School of Leisure, Sport and Tourism University of Technology, Sydney, pp 1–8

Gotham KF (2005) Theorizing urban spectacles. festivals, tourism and the transformation of urban space. City 9(2):225–246

Grant D, Paliwoda S (1998) Segmenting Alberta arts and festival consumers. Festival Manag Event Tour 5:207–220
Grant D, Paliwoda S (2001) Analyzing Alberta festival consumers. Event Manag 7:67–77
Griffin R (2012) Fundamentals of management. South Western Cengage Learning, Mason
Guest G, Namey EE, Mitchell ML (2013) Collecting qualitative data: a field manual for applied research. Sage Publications Inc., London
Gursoy D, Kim K, Uysal M (2003) Perceived impacts of festivals and special events by organizers: an extension and validation. Tour Manag 25(2):171–181
Hall CM (1989) The definition and analysis of hallmark tourist events. Geojournal 19(3):263–268
Hall CM (1992) Hallmark tourist events: impacts, management. Planning, Belhaven, London
Hartshorne R (1939) The nature of geography. Association of American Geographers, Lancaster, Penn
Iso-Ahola S (1982) Toward a social psychology theory of tourism motivation. Ann Tour Res 12:256–262
Isar RF (1976) Culture and the arts festival of the twentieth century. Cultures 3:125–145
Jago L, Chalip L, Brown G, Mules T, Ali S (2003) Building events in destination branding: insights from experts. Event Manag 8(1):3–24
Jago L, Dwyer L (2006) Economic evaluation of special events. A practitioner's guide, Common Ground Publishing Pty Ltd—CRC for Sustainable Tourism Pty Ltd, Altona Vic
Janeczko B, Mules T, Ritchie B (2002) Estimating the economic impacts of festivals and events: a research guide. CRC for Sustainable Tourism Pty Ltd, Gold Coast
Jones R, Pilgrim A, Thompson G, Macgregor C (2008) Assessing the environmental impacts of special events: examination of nine special events in Western Australia. Gold Coast, Queensland (Sustainable Tourism Cooperative Research Centre)
Johnston RJ, (2007) City and society, an outline for urban geography. Routledge, Abingon–Oxon
Johnston RJ, Gregory D, Smith DM (1996) The dictionary of human geography. Blackwell, Oxford
Kowalczyk A (2000) Geografia turyzmu. PWN, Warszawa
Krausse GH (1998) Waterfront festivals: a spectator analysis of event tourism in three New England cities. Festival Manag Event Tour 5:171–184
Kulshrestha UC, Nageswara Rao T, Azhaguvel S, Kulshrestha MJ (2004) Emissions and accumulation of metals in the atmosphere due to crackers and sparkles during Diwali festival in India. Atmos Environ 38:4421–4425
Leszczycki S (1972) Perspective on development of geographical sciences. Presidential address, XXII IGC, Montreal, August 11. IGU Bulletin 23(2)
Lee CK, Lee YK, Wicks BE (2004) Segmentation of festival motivation by nationality and satisfaction. Tour Manag 25(1):61–70
Lee I, Arcodia C, Lee TJ (2012) Benefits of visiting a multicultural festival: the case of South Korea. Tour Manag 33(2):334–340
Lee SY, Petrick JF, Crompton J (2007) The roles of quality and intermediary constructs in determining festival attendees' behavioral intention. J Travel Res 45(4):402–412
Leszek W (1977) Zasady eksperymentowania. Wydawnictwo Politechniki Poznańskiej, Poznań
Liszewski S (1995) Przestrzeń turystyczna. Turyzm 5:87–103
Long P (2000) After the event: perspectives on organizational partnership in the management of a themed festival year. Event Manag 6(1):45–59
Long P, Perdue R (1990) The economic impact of rural festivals and special events: assessing the spatial distribution of expenditures. J Travel Res 28(4):10–14
Macfarlane I, Jago L (2009) The role of brand equity in helping to evaluate the contribution of major events. CRC for Sustainable Tourism Pty Ltd, Gold Coast
Madden J, Groenewold N, Thapa P (2002) Estimating demand for a special event and the event's contribution to a regional economy. CRC for Sustainable Tourism Pty Ltd, Gold Coast
Markova B, Boruta T (2012) The potential of cultural events in the peripheral rural Jesenicko region. Acta Univ Carol Geogr 47(2):45–52

Marvell A, Sim D, Schaaf R, Harper R (2013) Students as scholars: evaluating student-led learning and teaching during fieldwork. J Geogr High Educ 37(4):547–566
McDowall S (2010) A comparison between Thai residents and non-residents in their motivations, performance evaluations and overall satisfaction with a domestic festival. J Vacation Market 16:217–233
Mohr K, Backman KF, Gahan LW, Backman SJ (1993) An investigation of festival motivations and event satisfaction by visitor type. Festival Manag Event Tour 1(3):89–97
Molloy J (2002) Regional festivals: a look at community support, the isolation factor and funding sources. J Tour Stud 13(2):2–15
Nicholson R, Pearce DG (2000) Who goes to events: a comparative analysis of the profile characteristics of visitors to four South Island events in New Zealand. J Vacation Market 6:236–253
Nicholson R, Pearce DG (2001) Why people attend events: a comparative analysis of visitor motivations at four South Island events. J Travel Res 39:449–460
Pacione M (2005) Urban geography: a global perspective. Routledge, London–New York
Pahl N, Richter A (2007) SWOT analysis—idea, methodology and practical information. GRIN Verlag, Norderstedt
Picard D, Robinson M (eds) (2006) Festivals, tourism and social change. Remaking worlds, Channel View Publications, Clevedon
Pieper J (1965) In tune with the world: a theory of festivity. Harcourt, New York
Piette A (1992) Play, reality, and fiction: toward a theoretical and methodological approach to the festival framework. Qual Sociol 15(1):37–52
Przybylska L (2015) Between sacralization and festivalization of public spaces: a case study of the Cavalcade of the Three Kings in Poland. Bull Geogr Socio-econ Ser 27:171–180
Quinn B (2003) Symbols, practices and myth making: culture perspectives on the Wexford Festival Opera. Tour Geogr 5(3):329–349
Quinn B (2005a) Arts festivals and the city. Urban Stud 42(5–6):927–943
Quinn B (2005b) Changing festival places: insights from Galway. Soc Cult Geogr 6 (2):1470–1197
Quinn B (2009) Festivals, events and tourism. Festivals, events and tourism. In: Jamal T, Robinson M (eds) The SAGE handbook of tourism studies. Sage, London, pp 483–503
Radclife SA (2006) Culture in development thinking: geographies, actors, and paradigms. In: Radclife SA (ed) Culture and development in a globalizing world: geographies, actors and paradigms. Routledge, pp 1–29
Rao V (2001) Celebration as social investments. Festival expenditures, unit price variation, and social status in rural India. J Dev Stud 38(1):71–97
Remmen KB, Frøyland M (2014) Implementation of guidelines for effective fieldwork designs: exploring learning activities, learning processes, and student engagement in the classroom and the field. Int Res Geogr Environ Educ 23(2):103–125
Ritchie J (1984) Assesing the impact of hallmark events: conceptual and research issues. J Travel Res 23(1):2–11
Robinson M, Picard D, Long P (2004) Introduction-festival tourism: producing, translating, and consuming expressions of culture(s). Event Manag 8(4):187–242
Runge J (2007) Metody badań w geografii społeczno-ekonomicznej. Wydawnictwo Uniwersytetu Śląskiego, Katowice
Ryan C, Gu H (2010) Constructionism and culture in research: understandings of the fourth Buddhist Festival, Wutaishan, China. Tour Manag 31(2):167–178
Saayman A, Saayman M (2006) Sociodemographics and visiting patterns of arts festivals in South Africa. Event Manag 9(4):211–222
Saleh F, Ryan C (1993) Jazz and knitwear: factors that attract tourists to festivals. Tour Manag 14 (4):289–297
Saleh F, Wood C (1998) Motives of volunteers in multicultural events: the case of Saskatoon folk fest. Festival Manag Event Tour 5:59–70

Smith A, Fox T (2007) From 'event-led' to 'event-themed' regeneration: the 2002 Commonwealth Games Legacy Programme. Urban Stud 44(5/6):1125–1143

Schneider IE, Backman SJ (1996) Cross-cultural equivalence of festival motivations: a study in Jordan. Festival Manag Event Tour 4(3–4):139–144

Turner V (ed) (1982) Celebration: studies in festivity and ritual. Smithsonian Institution Press, Washington

Urban S, Ładoński W (2001) Jak napisać dobrą pracę magisterską. Akademia Ekonomiczna im Oskara Langego we Wrocławiu, Wrocław

Uysal M, Gahan L, Martin B (1993) An examination of event motivations: a case study. Festival Manag Event Tour 1(1):5–10

Van Vinkle CM, Backman K (2009) Examining visitor mindfulness at a cultural event. Event Manag 12(3–4):163–169

Van Zyl C, Botha C (2004) Motivational factors of local residents to attend the Aardklop National Arts Festival. Event Manag 8(4):213–222

Veal AJ (2006) Research methods for leisure and tourism: a practical guide. Pearson Education Limited, Harlow

Visser G (2005) Let's be festive: exploratory notes on festival, tourism in South Africa. Urban Forum 16(2–3):155–175

Warszyńska J, Jackowski A (1978) Podstawy geografii turyzmu. PWN, Warszawa

Waterman S (1998) Carnival for elites? The cultural politics of arts festivals. Prog Hum Geogr 22 (1):54–74

Wood E, Rhodri T (2006) Measuring cultural values—the case of residents' attitudes to the Saltaire Festival. Tour Econ 12(1):137–145

Wrona J, Rek J (1998) Podstawy geografii ekonomicznej. Polskie Wydawnictwo Ekonomiczne, Warszawa

Yuan J, Cai LA, Morrison AM, Linton S (2004) An analysis of wine festival attendances' motivations: a synergy of wine, travel and special events? J Vacation Market 11:41–56

http://www.pwn.pl. Online Encyclopedia of Polskie Wydawnictwo Naukowe. Accessed 20 Feb 2015

http://komiksfestiwal.com/en/—Website of International Festival of Comics and Games in Łódź. Accessed 01 March February 2015

Chapter 4
Festivalisation and Its Effects on the Urban Space

Abstract In this chapter, the basic conceptions and definitions related to festivalisation, the main impacts of festivals on space, the concept of festival tourism and the most important types of festivalisation spaces are characterised. The description of the basic conceptions and definitions related to festivalisation includes concepts such as geographical space and its types, or urban space. This is the theoretical basis for the analysis developed further in the book. The concept of festivalisation and its main features are investigated in this chapter as well. In the next subchapter, the most important impacts of festivals on urban space, divided into tangible, intangible, positive, and negative ones, are described. Further on, the reader can find references to festival tourism, first to the understanding of the concept of tourism and its classifications. The definitions of tourism, tourists and tourist assets are given here as well. Later the definition of festival tourism as one of the tourism types is formulated, and its influence on tourism space is briefly presented. The last subchapter presents festivalisation spaces. They are urban spaces which festivals influence most strongly: e.g. festival centres as spaces of permanent festivalisation, or theatres, museums, streets and squares, occupied temporarily for festival purposes. The theoretical characteristic of different festivalisation spaces is supplemented with many practical examples.

Keywords Festivalisation · Urban space · Urban space dimensions · Festivals impacts · Festivalisation spaces · Types of festivalisation spaces

4.1 Basic Concepts

For the purpose of this chapter, it is essential to specify the concept of space and the understanding of it in geography and in urban studies. Space is a fundamental concept in geographical sciences; thus, it was decided to show how space is understood in geography first. Geography is a science where spatial analysis is the basic research tool. It started to develop in ancient times, when it was treated as a description of the Earth's space (Greek: hê gê—'Earth' and graphein—'to write'—see Geographica by

W. Cudny, *Festivalisation of Urban Spaces*, Springer Geography,
DOI 10.1007/978-3-319-31997-1_4

Eratostenes, Leszczycki 1975). As regards the definition of space itself, physics defines it as "a boundless, three-dimensional extent, in which objects and events occur and have relative position and direction" (http://www.britannica.com/EBchecked/topic/557313/space). However, in geography, now it is not possible to define space in such a simple way, due to the multitude of aspects and the complexity of geographical studies. With time, the idea of a relational (abstract, relative and subjective) space gained in popularity. The object of study in geography is the three-dimensional geographical space (sometimes referred to as Earth's space or Earth's cover, called epigeosphere), filled with objects, phenomena and processes. What is important, however, is that these processes, phenomena and relations often assume an intangible, relative, relational dimension (Lisowski 2003). According to Leszczycki (1975), the study of location of such phenomena and spatial relations among them is the object of geography. As Thrift (2003, p. 95) wrote, "… space is not a commonsense external background to human and social action. Rather, it is the outcome of a series of highly problematic temporary settlements that divide and connect things up into different kinds of collectives which are slowly provided with the means which render them durable and sustainable" (for debate on geographical space, see also: Siwek 2011). According to Thrift (2003 cited in Cudny 2014, p. 138), there are four dimensions of space investigated in geography:

1. The first space—empirical constructions: visible, measurable elements, physically present in our surroundings.
2. The second space—the unblocking space: it consists of connections (social and economic) through which we know how the world interacts.
3. The third space—the image space: it consists of pictures or images. Images of different kind are exemplifications of space (pictures, photographs, satellite images, movies, etc.) and they are studied in geography.
4. The fourth space—the place space: here space is understood as a place. A place "…consists of particular rhythms of being that confirm and naturalize the existence of certain spaces… For places not only offer resources of many different kinds (for example, spatial layouts which may allow certain kinds of interaction rather than others) but they also provide cues to memory and behavior".

Following Kostrowicki's classification (1997), geographical space can also be divided into objective and subjective space, depending on its measurability, perceptibility and values. The former is a space which is easily measured and perceived (geometrical, physical and ecological), while the latter is the result of the human perception of the surroundings, so it depends on individual, subjective (ethical, psychological or cultural) factors. Lisowski (2003) divided geographical space a little differently, from the human geography perspective. In his work regarding the role of space in human geography research, Lisowski (2003 cited in Cudny 2014, p. 138) identified four basic types of space:

1. Physical space, treated as a set of physical objects on the Earth's surface.
2. Ecological space, treated as the life space surrounding an individual. It includes both natural and technological–economic elements.
3. Cultural space: it is a physical space which is a particular carrier of meanings (cultural landscape) or may take the form of a formal, abstract space (a system of signs).
4. Social space, which this author derives from the works by Henri Lefebvre and Manuel Castells, the sociologists. In this case, space is treated as an element produced by the society and expressing social relations. Lisowski claims (after Manuel Castells) that this type of space includes the economic, cultural and political dimensions, as well as (after Henri Lefebvre) the abstract and concrete subspaces.

As regards the definition of urban space, it may be found, e.g., in the publication by Liszewski (2008 p. 189). He claims that "urban space is a part (subspace, partial space) of geographical space, which is characterised by a particular organisation and landscape, with dominant non-agricultural human activity, and this area has a formally defined legal status (of a city). The space is inhabited by a distinctive local community".

When discussing the influence of festivals on urban space, one of the most important tasks is to define the concept of festivalisation. As it was pointed out earlier, members of the modern experience societies are increasingly searching for varied experiences. Nowadays, people form extremely mobile and modern societies. In the surrounding world, transport and communication are becoming very easy and fast. We are slowly becoming a part of mobile life, with fast transport, mobile phones, mobile Internet or television. This sometimes makes us do many different activities at the same time (Elliott and Urry 2010). Human mobility and the speed of contemporary life are also reflected in culture. Currently, festivals are becoming its increasingly popular element. They meet the requirements of contemporary societies as they are dynamic events, remaining in continuous motion. They often take place at different locations, encompass various arts and create conditions for direct encounters with artists. The development of festivals and their growing role in culture are referred to as culture festivalisation (Hauptfleisch et al. 2007).

Apart from culture, the process of festivalisation (German: Festivalisierung) regards other spheres of human life as well. According to Richards and Palmer (2010, p. 28), festivalisation encompasses a temporary transformation of the city into a particular, symbolic space, where the use of the public sphere is affected by a specific model of taking advantage of culture (Van Elderen 1997). On the other hand, Roche (2011, p. 127) claims that "Festivalisation can be taken to refer to the role and influence of festivals on the societies that host and stage them—both direct and indirect, and in both the short and the longer term". Festivalisation is a process which has many consequences, which impacts on broadly understood urban space, in both its objective and its subjective dimension. Festivalisation is used for the development of cities, improvement of their image and gaining larger income.

It may also be a way to revitalise and restructure neglected urban centres. Moreover, it performs many social and cultural functions with regard to city inhabitants and the visiting tourists.

Summarising in the case of cities, festivalisation should be treated as a complex process of the development of festivals and their influence on various elements of urban space. This influence may be divided into tangible and intangible and positive and negative impacts. The tangible impacts of festival development are quantifiable and include the value of investment, income and employment generated by different activities connected with the festival event. Intangible impacts are not easily quantifiable; they are connected with the non-material sphere of perception, education, social connections, etc. For example, overcrowding and inconvenience to local residents due to the organisation of events could be one of them. Intangible benefits include an increased social capital and business experience, or development of knowledge, community cohesion and pride, development of non-material culture or creation of a positive image of the city (Dwyer et al. 2000; Carlsen 2009).

4.2 The Main Impacts of Festivals

One of the aims of this book is to present the most important spheres and types of festival impacts on urban space (for the analysis of festival's impacts, see also Cudny 2013, 2014; Cudny et al. 2012a, b; Davies 2015). Another aim is to evaluate this influence, i.e. establish the positive and negative effects of festivals. Referring to the concepts of geographical space and festivalisation, described in Sect. 4.1, we can also distinguish different elements of urban space, both objective and subjective (Table 4.1). In the process of city festivalisation, festivals have a direct and indirect influence on all these elements of urban space.

It should be stressed that similar to many other phenomena, festivals have a number of positive and negative effects on the surroundings. Among the impacts broadly described in the literature on the subject (e.g. in event studies), there are the economic effects of festival activity. In this case, a festival generates tourism and consequently becomes a source of income whose beneficiaries are local firms

Table 4.1 Different elements of urban geographical space

Objective	Subjective
Physical space—empirical constructions	Unblocking space—socio-economic linkages within space
	Third space—image space
Place—space	
Ecological space	Cultural space
	Social space—including economic, cultural and political elements

Source Author's compilation

providing services for the event. Analyses of the economic impacts of festivals are quite common in contemporary studies (see Chap. 3). They indicate that these impacts may be direct or indirect. In the latter case, we are talking about the multiplier effect.

However, we must not forget about the negative economic effects of festivals like the high cost of organising them. The expenditures include artists' remuneration, the cost of preparing the stage, securing proper safety measures, energy and water supply, and waste disposal. Events, especially mega events, are so expensive that the profits gained from them do not compensate for the costs. Moreover, in many cases, festival organisers overstate the number of visitors or overestimate the expected profits, in order to, for instance, secure the local authorities' favour in future. Another important problem is the fact that local or regional firms are often ousted by global companies from servicing festivals. This in turn limits the positive influence of the event on the local economy (Davies 2015, pp. 551–552). There is also the question of taxing the organisers, artists and firms providing services during mega festival events. Analyses concerning mega sports events show that many firms and institutions providing services are granted substantial tax relief, which may result in lower budget revenues (see Tetłak 2013, 2014). It seems that an interesting move, also from the practical point of view, would be to start research in this area, with reference to large festivals.

It should be stressed that there are many festival impacts which are not related to economy. Davies (2015, p. 555) presented the main positive and negative features of festivals as their advantages and disadvantages. The author demonstrated that festivals have an effect on people as well, as regards both individuals (excitement, pleasure, increased local pride due to participation in the festival) and whole groups (community participation/cohesion). He believes that festivals also have an influence on culture (improving culture, downgrading old cultures, etc.), people's safety (festival participants), generating new jobs, etc. It should be stressed again that all these complex and multifaceted impacts may have both a positive and a negative character.

This subchapter presents the issue of positive and negative festival impacts. However, they are related to the elements of the urban space (see Table 4.1) which they directly affect. The presentation was divided into impacts concerning physical space and place space (as the most measurable), ecological space, urban communities and economy (unblocking and social space), image and cultural space.

4.2.1 *The Influence of Festivals on Physical and Place Space*

Positive Impacts

Festivals are in many ways connected with physical space and place space. Firstly, they are based on physical space but also transform and create it, e.g. due to construction of large buildings for festival purposes. Impacts on the objective space should be regarded as tangible effects of festivalisation. Examples of objects built specially for the purpose of huge film festivals are the Palais des Festivals in Cannes or the Berlinale Palast in Berlin (de Valck 2007). Festivals also influence urban space through microchanges, e.g. building temporary facilities, such as concert stages or the stalls, or marking temporary borders of roads and squares where the events are held.

There are new buildings appearing next to permanent culture and entertainment facilities used during festivals, while the existing ones are often revitalised. In this way, large parts of the city may be renewed. It should be stressed that erecting such buildings is often a part of purposefully implemented programmes of revitalisation through culture. They involve not only creating non-material cultural assets, but also investing in erecting new buildings used for service and organisation purposes. New investments or revitalised areas are accompanied by newly developed elements of permanent technical infrastructure, e.g. roads, telecommunications and power production devices. This is also a positive impact of festivalisation on the physical and place space of urban areas.

Another positive impact is the influence on the objective space in the physical dimension, as a part of the so-called urban acupuncture. It involves introducing slight changes at particular points of the urban tissue. Such changes include erecting or renovating individual buildings or creating small urban spaces. They must be located at nodal sites, the meeting points of inhabitants, transport systems and natural environment. With the help of small investments, they are to initiate further, in a way spontaneous positive transformation of cities. They are to be a kind of a nucleus around which further investments will be made—buildings and new revitalised spaces (see Lerner 2003). "A festival based on a particular theme, where interventions are strategically concentrated on specific areas of a city, can be considered a temporary Public Space Acupuncture strategy. They also modified citizens' perceptions of their own city, reinforcing the emotional relations between people and their public space (Casanova and Hernandez 2011)".

Negative Impacts

Festivals may also have a negative influence on both physical space and place space in urban areas. They may, for instance, lead to the destruction of space, also historical (heritage), by developing it in the wrong direction or by generating acts of vandalism by festival participants. They temporarily disturb the functioning of space by excluding a part of it from normal use (e.g. streets, squares, buildings). Moreover, some buildings may be erected at the wrong places, have wrong technical parameters or simply be ugly, unattractive, spoiling physical space and the cityscape (city physiognomy) and unaccepted in the space, e.g. by the inhabitants.

4.2.2 The Role of Festivals in Shaping the Ecological Space

Positive Impacts
As regards the natural environment, the development of events and festivals exerts tangible and intangible impacts on the objective space. Some of them are visible and measurable, e.g. thanks to devices measuring pollution or costs of utilising waste. Some effects do not show the quality of measurability, which mainly concerns the positive influence (Fredline et al. 2006). The advantages include the fact that it is possible to present ideas related to environmental protection and sustainable development through festivals (Cudny 2013). As a result, it is possible to popularise and develop them among the society.

Negative Impacts
Negative impacts include first of all the degradation of the environment caused by air and water pollution, noise, littering and overcrowding (Arcodia and Cohen 2007). The negative effects of events result mainly from the inflow of festival visitors into the host destinations. Visitors often travel by car, which pollutes the atmosphere. The greater number of people staying in the city causes an increased consumption of electricity (and automatically a larger emission of CO_2) and water (which increases the amount of sewage) and produces more litter. Moreover, when festivals are held in the open air, the area may get destroyed (e.g. the vegetation) and littered as a result of crowds gathering at one place. These are measurable effects, so they must be included in tangible impacts. It should be stressed that in the developed countries (e.g. Australia), the analysis of the environmental impacts of festivals is more and more often used in festival management (Schlenker et al. 2010, p. 8; see also Cudny 2013). Larger traffic resulting from festivals increases the level of troublesome noise. Festival facilities visited by tourists and city inhabitants are crowded, which may also be considered to be a negative element. An interesting example of a negative influence of festivals on the environment is the use of fireworks, for example, during big festivals in India (Cudny 2013).

4.2.3 The Role of Festivals in Forming the Unblocking Space and Social Space

Positive Impacts
Festivals are an element which strongly influences the urban community, which forms the unblocking and social urban space. First of all, they are a component which creates a broadly understood social capital. It is constituted by interpersonal linkages, norms and social relationships. Festivals reinforce the social capital because they build relationships between their organisers and the inhabitants, local entrepreneurs, politicians, volunteers and the festival staff. Moreover, they enable members of a given community to spend their free time and celebrate together

(Arcodia and Whitford 2006). Festivals are also an arena for expressing and consolidating opinions and values common for the urban community (Marston 1989).

Festivals are a form of organising free time for the city inhabitants and the tourists visiting the city (Cudny and Rouba 2011). Urban festival spaces offer creative possibilities because they temporarily suspend social relations. They also sustain playful practice that may challenge established society norms (Waitt 2008). This makes festivals a platform for interesting and close interpersonal relations. They are places where one can get away from the daily routine; attending them is a form of rest and relaxation and a way of spending time with the family, strengthening the family ties. We may also develop our relations with friends and meet new, interesting people. These last two elements are a strong social motivation of festival attendees, which was confirmed during many events (Lee et al. 2004). In this sense, the free time spent at festivals is also an element shaping the social capital within the family and among friends.

In recent years, we have observed the formation of creative societies, based on knowledge and innovativeness. Expanding (accumulating) knowledge and skills, which are a part of human capital, is extremely important. Creativity is an element typical of cities, where three basic elements determining their development are found: technology, talent and tolerance (Florida 2003). Cities are where technological development centres (universities laboratories, high-tech industry) operate and where we can find educated, talented people working at these centres, who Florida refers to as the creative class. It is also where events broadening human horizons take place, building the foundation of tolerance and acceptance of diversity, such as festivals. In this context, festivals are an extremely interesting element developing human knowledge. Thanks to them, we broaden our mental horizons, discover new cultures and gain knowledge about travel (e.g. at travel festivals) or science and technology (Cudny and Rouba 2011).

Various events, including festivals, make it possible to develop and promote culture. They also stimulate older and young people to participate in culture by taking part in them (Derrett 2003; Mikkonen and Pasanen 2010). Moreover, in the era of multicultural societies, consisting of different groups of migrant population, they are an element supporting multiculturality and developing tolerance attitudes. It is possible due to presenting the immigrant culture during these events and the participation and involvement of people from outside these groups (Lee et al. 2010; Buch et al. 2011). They are also a meeting place for sexual minorities, e.g. gay and lesbian, an arena where they can publicly manifest their sexual orientation (Markwell and Waitt 2009). It may be treated as a positive element from the social point of view, though, on the other hand, some of such festivals may raise objections on the part of some inhabitants.

Negative Impacts

According to Waitt (2008), festivals may be socially burdensome as they cause certain social helplessness. It occurs when festivals are an oppressive social force excluding some people from or including them into public space. An obviously significant social problem in the functioning of some festivals is the exclusion of a part of the inhabitants (e.g. the poor) from the participation in festivals (Cudny 2013). The social cost of organising festivals in cities is also the presence of conflicts between the organisers, authorities and other festival stakeholders. Conflicts affect many spheres of city life. They depreciate the brand of the city and destroy social capital by disturbing the networks of social linkages created around the festival (Cudny 2011). They cause social problems, e.g. by generating pressure among supporters of individual options and disagreeing about the organisation of the festival. Festivals often become the source of conflicts between the inhabitants and tourists, which is a negative thing. The conflicts result from the inconveniences caused by event organisation, such as the local inhabitants' impression of some kind of an alien invasion (felt especially strongly in small destinations), traffic jams, problems finding a parking space, crowds gathering around large festivals, noise and higher prices of services and goods (due to a larger demand) (Mikkonen and Pasanen 2010).

Festival studies often point to frequent abuse of alcohol and drugs, as well as risky sexual practices during festivals, which should be treated as negative impacts of festivals on the unblocking and social urban space (see Hughe et al. 1995; Holt 2005). Moreover, during some festivals, there are incidents of crimes, acts of vandalism and situations involving nudity and vulgar behaviour, which are very often highly disapproved of by some inhabitants of the host destinations (Gotham 2005; Cudny 2013).

4.2.4 *The Role of Festivals in Shaping the Economic Aspect of the Unblocking Space*

Positive Impacts

Fredline et al. (2006, p. 3) present the effects of events, including festivals, referring to Ritchie (1984) and Hall (1989, 1992). The economic impacts include generating income by smaller events within a given community. It is a kind of money redistribution, resulting from the fact that local festivals that do not attract large numbers of tourists are not financed from sources outside the region/city. In the case of larger events, generating significant tourist traffic, such as large film festivals (Cannes, Berlin, Venice), the money flows from outside. The festivals are attended by tourists who must buy goods and services on the spot (food, fuel, hotel services, etc.). In this way, a given destination receives money from other areas, developing its economy in this way. In addition, new jobs in festival management are generated, and firms working for the events are developing, e.g. restaurants, catering firms, hotels, taxi companies, and promotion and advertising companies. Apart from that, festivals can prolong the tourist season considerably by attracting tourists, and in this way, they generate additional income for the city (de Valck 2007, p. 76). This kind of influence is economic in nature and should be regarded as an impact on the unblocking space in its economic aspect.

With the economic impacts presented above, many post-industrial cities nowadays invest in the development of culture. It is to be a remedy for the economic crisis after the collapse of industry. Investing in facilities where cultural activity is to take place and in a variety of events is becoming an element of far-reaching plans of the revitalisation and functional–economic restructuring of cities in Europe and outside it (Miles 2005; McCarthy 2006; Pratt 2009). Restructuring through culture usually consists of two main elements: erecting spectacular buildings connected with culture to make the city famous and attractive (e.g. the Guggenheim Museum in Bilbao, or the Berlinale Palast in Berlin) and creating experiences connected with a given destination, including festivals (Pratt 2009, p. 1042). As shown by research conducted in different countries, they have a significant influence on the economic sphere of regions and cities (Ryan 1998; Saayman and Saayman 2004; Tomho 2005; Pasanen et al. 2009). Effects of this kind should be included in tangible impacts. According to Hughes (1999) and Atkinson (2005), festivals may also be an element promoting a given destination and based on its heritage. In this way, the image of the city improves in the eyes of the tourists, investors and even inhabitants themselves. This may bring positive economic impacts, such as increasing the tourist traffic or investments. During the system transformation in the countries of Central and Eastern Europe, culture (including festivals) became an element used for the economic transformation of cities in post-communist countries (see Cudny et al. 2012a, b). Currently, we can observe rapid development of festival activity in this region as well.

In recent years, the importance of creative thinking and acting has grown rapidly. At present, it is the basis of social and economic development of modern countries, regions and cities (Florida 2003). Creativity is understood as the foundation of innovation, and innovation is the new primary driver of economic growth and the basis for the development of post-industrial new economy (Flew 2010). Creativity is reflected in the development of creative industries, which are an important type of activity in modern cities. This sector includes all activities engaging skills and talent in order to create a new, original work or product, which is basic to generating prosperity and jobs for the present and the future, using the rule of intellectual property. The following activities (industrial and service) could be included into the creative industry field: advertising, architecture, arts and antique markets, crafts, design, designer fashion, film, interactive leisure software, music, television and radio, performing arts, publishing and software (Creative Industries Task Force Document 2001 cited in Cunningham 2002, p. 1). Creative industries are a multidimensional sector encompassing activities connected with broadly understood creation of works (e.g. pieces of art), services (e.g. advertising) and cultural experiences (e.g. related to the cinema, art, museums) (Thomas et al. 2010).

The compilation above shows that cultural and artistic activity is also included in creative industries, which results from a large amount of the artists' own creative work, individually created new conceptions and ideas, as well as a considerable share of intellectual capital (Caves 2000). In this case, also festivals may be treated as an important component of creative industries. As Rofe and Woosnam (2015, p. 4) stated, "Previously seen as temporary spectacles, events and festivals have been recast within the policy arena as creative clusters... Festivals offering moments of creative participation and engagement that contribute to the vibrant culture of the city. Herein rests the circular logic of the creative city ethos; that the fostering of creativity in place attracts creatives and creatives themselves infuse place with creativity". For instance, during film festivals, their participants may practise at workshops conducted by renowned professionals. The latest film-making techniques and equipment are presented as well. Moreover, film professionals, i.e. actors, film directors and cinematographers, have direct contact with film producers and agencies. A film festival is then a place and a time of creative people and ideas, so important for the economic development of cities, coming together (Mezias et al. 2011). The effects of festival development described above should be regarded as economic intangible impacts, belonging to the unblocking space.

Negative Impacts
On the other hand, according to Fredline et al. (2006, p. 3), events also generate negative economic effects, e.g. considerable costs of tangible nature. They include using public financial resources, e.g. from the state, regional or city budget, which could be used for other purposes. Another disadvantage is the constant growth of prices of services and goods, resulting from the increasing demand generated by the tourists arriving at a given destination. The negative economic impacts also include the lack of control of public subsidies, i.e. for art festivals, paid from the city budget (Frey 1986). It must be said that in some publications, the economic impacts of festivals and events are overestimated. It is caused either by improper research methods or by the subconscious attempts to meet the expectations of organisers and sponsors, who demand large profits (Crompton and McKay 1994). Some authors also point to the fact that activity connected with art and culture alone is not enough to successfully replace the declining industrial function and overcome the resulting social problems in post-industrial cities. It may just be an element supplementing other activities aiming at restructuring cities economically (Miles 2005). Impacts of this kind are examples of how, e.g., festivals may burden the economic sphere of urban space.

4.2.5 The Role of Festivals in Creating an Image Space

Positive Impacts
Image of space is an important element of geographical cognition, and there is a lot of research devoted to the perception (image) of space in geography (see Siwek 2011). Culture and festivals take active part in the process of image creation in the context of geographical space. Creating a positive image or, to put it more broadly, promoting cities through cultural events is an intangible impact (Cudny 2014). However, consolidating information about a given city and causing it to be favourably received may result in tangible outcomes as well, e.g. economic ones.

In the 1980s, scientific literature presented a discussion concerning city marketing (also called territorial marketing). It became an indispensable element of urban strategy, which allows the cities to compete with one another on the global arena (see de Valck 2007; Andersson 2014). An extremely important function of festivals and other cultural events is, e.g., creating the image of host cities. Festivals create the brand of a given destination, which may have the features of a product brand. In this case, the city becomes a kind of an attractive place product (Ooi 2002a), e.g. a tourist product, an investment product (place)—an area attracting investors and a product (place) attracting immigrants and supporting local inhabitants (Ooi 2002a; Dinnie 2011). The image of a given city, created also through culture and festivals, as well as its brand, influences all these elements, i.e. the tourists, inhabitants and investors searching for new locations.

According to Jensen (2007, p. 212–213), "The etymology of the word *branding* literally implies the notion of burning, but we have left behind the notion of burning cattle and are now dealing with burning the consumer mind". According to Gnoth (2004), "Destinations develop their own brands very much for the same reasons as the manufacturing sector. They help identify and distinguish the company's offering over time and represent both a promise and experience for current and future products". As regards the definition of brand, Kotler and Armstrong (2010, p. 259) state that "A brand represents everything that a product or service means to consumers. As such, brands are valuable assets to a company". Obviously, there are many differences between creating the brand of an industrial product and creating the brand of place (e.g. a city). Place branding includes not only creating the name, symbol or logo which identifies and distinguishes a given destination, but also creating an impression of a memorable journey to a given city, generating pleasant experiences and memories connected with it in order to persuade the recipient of the marketing message to choose this destination and not another. Creating the brand of a given place is a complicated process due to the fact that it has a heritage, formed by its history and population. All this must be taken into account in the complicated process of territorial marketing. According to Ooi (2010, p. 5), city branding is based on four basic principles: the city brand always presents only positive aspects of the place, it modifies public perception of the city, it is based on the city identity (heritage), and it provides a framework for locals and non-residents to imagine and experience the place.

Moreover, a brand is cocreated by various stakeholders, such as local politicians and officials, members of non-governmental organisations or businessmen (Govers and Go 2009). Each of these groups tries to exert the biggest influence possible on the destination brand, and their cooperation is a particular game of conflicting interests (Ooi 2002b). Modern city branding based on culture is very often also based on narration. This special story about the city presents the most significant and the most positive elements of its heritage. The aim of marketing narration is to produce the best possible image of the city and root it in the recipients' awareness. An important element of such narration used in territorial marketing are festivals. Their large number and variety have a positive effect on the city brand. They show the city as an interesting, diverse, cultural, cheerful and modern place. Besides, thanks to festivals, the city is often presented in the media commenting on these events. It may also be promoted by celebrities (e.g. actors, musicians) taking part in festivals. In this way, we gain airtime for presenting the destination, which usually does not have to be paid for. Profits from this kind of promotion can be, however, precisely evaluated, by calculating the advertising value equivalent mentioned earlier in the book, i.e. measuring the amount of time in the media dedicated to the presentation of the city in the context of a given event and calculating how much it would cost to buy comparable commercial time in a given station (e.g. TV station).

Negative Impacts

As regards creating the city brand, we may also observe some negative impacts of festivals on the image space. Firstly, the large number of interest groups that take part in this process may lead to conflicts, sometimes seriously damaging the city brand (Ooi and Pedersen 2010; Cudny 2011). Some studies also point to the possibility of a negative impact of festivals on the tourist image of the city, which happens in the case of a badly run promotional campaign and the poor quality of the event (Boo and Buser 2006). Other negative impacts of festivals on the image space include the danger of the lack of authenticity in the case of brands which have no foundation in the real heritage of the place. Apart from that, creating a brand may destroy the authenticity of the city due to its attempts to respond to the tourist demand. In such cases, we may observe a significant dissonance between the perception of the city by the tourists and its perception by the inhabitants (Ooi 2010).

4.2.6 The Role of Festivals in Culture—Shaping Cultural Space

Positive Impacts

Thanks to festivals, it is possible to achieve better development of urban culture understood as artistic and intellectual activities. This type of effects can be included among the intangible impacts on the subjective geographical space of the city. It is possible to achieve by giving artists space for presenting their creations to the wide public. During festivals, artists may contact one another, which enables them to exchange artistic experiences and develop new creative ideas. The festival is also a platform for supporting young artists and promoting them. The organisers often hold competitions during which young artists fight for prestigious prizes, also financial. Festivals very often feature workshops at which experienced artists teach their trade to younger artists. The festival is also a forum for artists and the world of business, which is often the recipient of some types of art. Festivals are a common form of contact with culture and an interesting option of spending free time for the majority of city inhabitants (Cudny and Rouba 2011). Many event organisers offer free concerts and exhibitions, which often take place in easily accessible public spaces, such as a mall, street or city square. Festivals may be a place of popularising high culture and developing the so-called cultural capital (see Cudny 2011). For some organisers, it may even be one of the festival objectives. Cultural capital consists of several elements, such as tastes, familiarity with high culture, social and cultural conventions, formal education and culture-related property. A person gains cultural capital all his/her life through education, contact with culture and art, etc. (Bourdieu 1984).

Negative Impacts

According to Zukin (2010), contemporary cities more and more often lose their authenticity as a result of the progressing globalisation and the activity of institutions focused on maximising consumption. According to McClinchey (2008), cultural festivals are often planned as a part of place marketing, image branding and civic boosterism to aid in the regeneration of urban areas. In order to perform their functions better, they must adjust their offer to the needs of their most important recipients—mainly tourists. Prentice and Andersen (2003, pp. 26–27) believe that in this context, also festivals may cause a false perception of the local culture due to its trivialisation and adjusting to the tastes of mass recipient (commercialisation) and lack of authenticity. Sometimes festivals become a kind of mask concealing real urban problems, such as poverty, unemployment, lack of investment and social inequalities, behind a facade of fun (Gotham 2005).

Some people may have limited access to festivals due to the closed distribution or high prices of tickets. Moreover, sometimes only a small part of the tickets are sold on the free market and the majority of them are distributed, e.g., by organisers at their own discretion (Frey 1986). Tickets for some festivals are very expensive, which means that less affluent persons cannot afford to buy them.

4.2.7 The Roles of Festivals in Politics—Shaping the Unblocking Space or Social Space in the Political Dimension

Positive Impacts

Another intangible impact of organising urban festivals on the subjective space is that they play a political role (Cudny 2014). Festivals and other public events have been used for political purposes for decades. For instance, after the Second World War, political aims inspired the organisation of Berlinale—Berlin International Film Festival (at that time held in West Berlin). It was an element of the plan for reconstruction and democratisation of life in Germany and Europe after the fall of Hitler (Mezias et al. 2008). Film festivals are an element of building an image of a given country through culture. For example, an element building the image of Germany as a world centre of film culture is the rapid development (using, e.g., public funds) of the Berlinale festival in last years.

According to Frey (1994, p. 32), “Politicians form one of the groups benefiting directly from festivals and that is why they are interested in their existence and growth. They can be, for example, patrons of art festivals, subsidising them with public money. Thus, they participate in the most important festival events, like the opening gala, and may benefit from the accompanying media attention. It is especially important when there are well known celebrities participating, which always attracts extraordinary media attention". The above regards politicians at different levels, but this mainly concerns politicians working for the benefit of a given city. Allocating funds to festivals held in a given city and supporting investments connected with them is an element of political promotion.

Negative Impacts
Just like in the case of the festival impacts described earlier, also their political roles cause many problematic situations. For instance, such events were used for the purpose of regime propaganda already before the war, e.g. in Nazi Germany (Hagen 2008). Film festivals presented movies promoting specific political ideas, e.g. in communist countries, where the cinema and the whole public sphere were controlled by the state (censorship). The situation was similar in Nazi Italy, in Mussolini's times (Mezias et al. 2008).

Moreover, local culture and community may be manipulated by politicians in order to achieve immediate goals during large cultural–artistic events (Boland 2010). Events of this kind may be organised for the purpose of achieving particular aims, e.g. winning support before local elections. The organisation of events involves many stakeholders. Good cooperation, compromising attitude and building positive relations with other co-organisers make it more efficient. The stakeholders include both public and private organisations, which often collaborate (Quinn 2010). A situation when one or a group of stakeholders have different aims and try to put their interests first at any cost and ignore the remaining parties inevitably leads to conflicts (also political). This results in serious problems in the organisation and functioning of a given event (Larson and Wikström 2001). Conflicts that appear lead to many negative consequences, such as the depreciation of the city image, destruction of the network of linkages created around a given festival, negative influence on the city inhabitants or even liquidation of the festival.

4.3 Festival Tourism as the Main Festivalisation Factor

Before we start the analysis of festival tourism, it is necessary to concentrate on the concept of tourism in general. Besides, it is important to define other concepts closely connected with the tourist aspect of festivalisation. They include definitions of tourist assets or tourist product, as well as their types. Tourism is an extremely complicated phenomenon of many meanings. World literature is full of different definitions, which are presented in scientific publications, as well as various reports and legal documents, e.g. drawn by international organisations concerned with this issue (Wall and Mathieson 2006). According to Ryan (2003) cited in Cudny (2012, p. 12), tourism is: "the demand for, and supply of accommodation and supportive services for those staying away from home, and the resultant patterns of (1) expenditure, income creation and employment that are created, (2) the social, cultural and environmental consequences that flow from visitation and (3) the psychological changes that result for both visitor and host". According to the WTO (World Tourism Organisation), it "encompasses all linkages and phenomena which occur due to a change of the place of stay and the resulting stay of persons, while for these people this new place of stay is neither a permanent place of residence nor a place of work" (Gaworecki 2007, p. 18). At the same time, for statistical purposes, the following technical definition of tourism formulated by the WTO is commonly accepted: "Tourism comprises the activities of persons travelling to and staying in places outside their usual environment for not more than one consecutive year for leisure, business and other purposes not related to the exercise of an activity remunerated from within the place visited". The persons referred to in the definition of tourism are called visitors. A visitor is described as "any person travelling to a place other than that of his/her usual environment for less than twelve months and whose main purpose of the trip is other than the exercise of an activity remunerated within the place visited" (Wall and Mathieson 2006, p. 13). The category of "visitors" consists of international and domestic visitors. Each of these two groups consists of excursionists (not staying overnight at a given destination) and tourists, who spend at least 24 h at a given destination (they stay overnight) (Gaworecki 2007, p. 16).

Tourism development is based on tourist attractions. They could be defined as "(…) all those elements of a 'nonhome' place that draw discretionary travellers away from their homes. They usually include landscapes to observe, activities to participate in, and experiences to remember. Yet it can sometimes be difficult to differentiate between attractions and non-attractions. Transportation (e.g. cruise liners), accommodation (e.g., resorts), and other services (e.g. restaurants) can themselves take on the attributes of an attraction, further complicating the distinction between various segments of the tourism industry. At times, even tourists themselves can become attractions" (Lew 1987, p. 554). Tourist attractions or assets could also be defined according to Kowalczyk (2001) as "(…) all elements of the natural and non-natural (cultural or anthropogenic) environment which are interesting to tourists and make a given site, locality or area attractive from the tourist point of view" (Cudny 2012, p. 13). Those tourist attractions could also be divided into different types: natural, non-natural, recreational, sightseeing and specialist (see Cudny 2012). It is also possible to identify two other types of tourist assets, i.e. site and event assets. Site assets occur at a given place and are related to it, e.g. climate, landscape or history. Event assets are related to an event, e.g. cultural—entertainment or sports event. They attract tourists and motivate them to travel (Kaczmarek et al. 2010).

In the understanding presented by Kowalczyk (2001), tourist assets or attractions are something else than objects used for the purpose of tourist service, which should be included among the devices and facilities of tourist service, called elements of tourist infrastructure. Tourists arrive somewhere in order to take advantage of various tourist assets (e.g. sea, mountains, climate, historical monuments, museums and festivals). However, they cannot stay at the target destination without satisfying the basic material and non-material needs. This can be done, thanks to the tourist infrastructure, understood by Kaczmarek et al. (2010) as "those elements which secure the proper functioning of the tourist destination". They include accommodation facilities (e.g. hotels, guest houses, motels), catering facilities (e.g. bars, restaurants, coffee shops), as well as a group of facilities and devices, called auxiliary facilities, i.e. all elements that tourists need, which were not listed earlier. They are, for instance, recreation and sports facilities (swimming pools, tennis courts, sports halls, gyms) and sports equipment rentals, but also shops, cinemas, theatres, concert halls, banks, etc. This group includes both facilities serving only tourists and those which are used by tourists and inhabitants of the given area together (Cudny 2012, pp. 14–15).

In the light of the definitions and classifications presented above, festivals are typical tourist assets (see Getz 1991, 2008; Aldskogius 1993; Quinn 2006, 2010). First of all, they are an element attracting tourists to a given destination, and in this way, they match the definition of tourist assets. From the tourism point of view, festivals could be defined as experiences to remember, or elements of cultural environment that draw tourists away from their homes. Festivals should be treated as events and cultural or anthropogenic tourist assets, as they are derived from human (anthropogenic) forms of culture. They could also be called sightseeing attractions, because they are connected with learning, knowledge, experience and watching different emanations of culture. To some extent, festivals may be incorporated into specialist tourist assets which require a higher level of knowledge and skills on the part of the spectators. It is, for example, applicable to some high-culture events like opera or ballet festivals.

The relation between festivals and tourism can also be seen in tourist infrastructure. Festivals generate tourist traffic whose participants must use tourist infrastructure facilities, such as hotels or restaurants. Facilities working exclusively for festivals are quite rare. However, at destinations that host large and well-known regular events, festival visitors are an important segment of the tourist infrastructure users. Besides, festival events are held at auxiliary facilities, e.g. in theatres, cinemas or culture centres. This infrastructure includes also buildings erected specially for the purpose of organising festivals, e.g. the Palais des Festivals in Cannes (France) or the Adelaide Festival Centre (USA).

Significant elements of tourism are tourist products, without which commercial tourism could not function today. A tourist product is a simple, single service or a complex (consisting of several complementary services) offer, which is sold to tourists on the tourism market. Currently, the discussion over tourist products more and more often concerns the whole experience gained by the tourist while using the tourist product (tourist product as experience). We may include here satisfaction with the product, or its educational and social effects. As regards the structure of the tourist product, in the classic version, it has three basic dimensions. The first one is the core product—the most important element, encompassing the most significant benefits of and reasons for purchase. The second dimension is the facilitating product, which includes services indispensable to make use of the core product (e.g. accommodation, catering and transport). The third dimension is the augmented product, including additional elements which make the given product more attractive and competitive in relation to other tourist products (Cooper and Hall 2008).

Furthermore, there are many types of the tourist product, e.g. material (product—object) and non-material (product—service). There is a simple product, e.g. a single service, or a complex product, consisting of several components—services, e.g. a package tourist event. Spatial tourist products can be divided into products—buildings (e.g. a museum), trails (e.g. a mountain tourist trail) or products—areas (e.g. a national park). Finally, we can distinguish a particular tourist product called an event. This group of products includes sports, business events and festivals (Kaczmarek et al. 2010). The tourist product—event—may be defined as a "planned and organised happening (or a number of happenings), which integrates simple products and tourist assets, focused on gaining benefits in the psychological, social and economic sphere" (Kaczmarek et al. 2010, p. 107).

In this context, festivals should be regarded as non-material tourist products belonging to the category of events (see among others Cudny 2006; Huang and Lian 2006; Kaczmarek et al. 2010; Cudny et al. 2011; Kang et al. 2014; Dimitrić and Bjeljac 2015). They offer specific non-material tourist services, illustrating the definition of the product—event. It may be a simple tourist product, based on the experience of a music band's performance, which is the core product. However, it is often accompanied by additional services (e.g. gastronomic or even connected with accommodation), thus becoming a complex product. Many music festivals organise camping sites, basic catering, security service and medical care for their spectators. Some festivals offer more sophisticated services, e.g. by logging on a website, we can get access to hotel packages offered by hotels cooperating with the organisers and offering lower prices. In this situation, we deal with a facilitating augmented tourist product.

The next important issue connected with tourism is its typology. Tourism is not a homogenous phenomenon, but a complex and a very complicated one. Gaworecki (2007) presented a number of tourism classifications, based on various criteria, such as the number of participants and trips, participants' age, length of stay, season, type of accommodation and transportation used during the trip, sociological aspects and tourists' behaviour during the trip (Table 4.2). An important classification of tourism is the one based on the main purpose of travel. The question is what the tourist's purpose of visiting a given region or city is. Sometimes a tourist journey can be a combination of many types of tourism. On the other hand, some journeys may belong to different types of tourism. For instance, a trip in order to visit a museum may be treated as a part of sightseeing, cultural or urban tourism. Depending on the main purpose of travel, we may distinguish the following main types of tourism: sightseeing, recreational, qualified, spa and wellness, business, ethnic, cultural, religious and pilgrimage, urban, culinary and event tourism (see Gaworecki 2007; Reisinger 2009; Cudny 2012). We may ask ourselves: what type of tourism is festival tourism?

Table 4.2 Selected tourism divisions

Criteria	Forms of tourism
Number of participants and trips organisation	• Individual • Group (organised)
Participants' age	• Youth • Senior
Length of stay	• Short term • Long term
Season	• Summer, winter • Peak of the season • Outside the season
Type of accommodation	• Hotel • Semi-hotel (e.g. campers, trailers, second homes)
Transportation	Rail, bus, car, airplane, boat, foot
Sociological aspect	• Luxury tourism—exclusive—traditional • Youth—senior—social
Behaviour during the journey	• Mass tourism • Rational tourism

Source Author's compilation based on Gaworecki (2007, pp. 21–22)

It is rational to classify tourism where the aim (the main one or one of the main ones) is to attend a festival as a part of cultural and event tourism. It is also justifiable to regard it as a form of sightseeing, urban, rural or special interest tourism. Certainly, such tourism may be treated as a separate type, i.e. festival tourism.

It should be pointed out that nowadays culture strongly attracts tourists (Tighe 1986). The literature on the subject often presents the concept of cultural tourism (Davies 1993; Stebbins 1996; Alzua et al. 1998; Du Cros and McKercher 2002; Richards 2007; Buczkowska 2008; Ivanovic 2008; Smith 2009). According to the dualistic definition of cultural tourism, formulated for the Association for Tourism and Leisure Education (ATLAS), it is first of all "the movement of persons to cultural attractions away from their normal place of residence, with the intention to gather new information and experiences to satisfy their cultural needs". Secondly, it is "all movements of persons to specific cultural attractions, such as heritage sites, artistic and cultural manifestations, arts and drama, outside their normal place of residence" (Richards 1996, p. 24). We also often come across the notion of heritage tourism (Nuryanti 1996; Poria et al. 2001; Dallen and Boyd 2003), which is close to the notion of cultural tourism (Alzua et al. 1998). The term "heritage tourism" refers to human heritage, which includes all material immobile resources (e.g. buildings, rivers, natural areas), material mobile resources (e.g. museums, archival documents), non-material elements, such as values, customs and lifestyles, as well as events such as festivals and artistic and cultural events (Dallen and Boyd 2003, p. 3). The definitions above clearly show that travelling to festivals may be a part of cultural and heritage tourism. It comes from the fact that they are artistic and cultural manifestations, arts and drama cultural attractions, as well as elements of intangible cultural heritage, which are the aims of tourist trips (see Bjeljac et al. 2015). We find several acknowledgements of such thesis in the research literature. Kang et al. (2014) pointed out that travelling to festivals may be a part of heritage tourism. Those authors analysed the case of South Korean festival held in Baudeogi which is based on South Korean tradition and folklore. They discovered that the main purpose of travel among the festival goers is the curiosity and interest in cultural heritage of South Korea. In other work regarding Asian festival held in Chinese destination Wutaishan, Ryan and Gu (2010) ascribe visiting this event to the cultural tourism.

It is justifiable to include tourist trips made in order to participate in festivals into event tourism, a type defined in event studies. In this case, the tourist's aim is to participate in various kinds of events, including festivals (Getz 2008, 2010). Issues connected with event studies, types of events and relations between events and festivals were discussed in the previous parts of the book, and therefore, it would not be elaborated here.

It is also possible to include travelling to festivals in sightseeing, urban or special interest tourism. A festival may be a cognitive element, related to discovering, learning and visiting. It may be attended on the occasion of visiting the city or the rural area (food and wine, and rural or urban festivals). Moreover, it may also be a festival connected with the tourist's hobby or interests. However, in this case, the main aim of the journey is controversial, as it is not the festival, but a historical monument, a city, a rural area and its culture or environment that raises a person's interest or encourages them to take up a hobby.

Another type of tourism mentioned in the literature on the subject is festival tourism (see Long and Robinson 2004; Buczkowska 2008). It is often regarded as a subtype of the types of tourism described above, including cultural or event tourism. However, festival tourism should be treated separately, because of the distinctiveness of festivals as unique culture and entertainment events (see Chap. 2). Cudny (2013) presented the concept of festival tourism and festival tourists, as well as the influence of this type of tourism on tourism space. According to him, festival tourism should be treated as a separate type of tourism, where the main aim of the journey is to take part in a festival. Tourists choosing this type of tourism should be referred to as festival tourists.

The previous subchapter presented the positive and negative impacts of festivals on different elements of urban geographical space. At this point, this topic would be briefly referred once again, this time in the context of tourism functions and dysfunctions. The analysis of tourism impacts, both general and referring to its individual types, is often conducted by dividing them into positive outcomes (functions) and negative outcomes (dysfunctions). This type of analysis was presented among others in the publications by Szwichtenberg (2000) and Gaworecki (2007). This idea was also developed by Cudny (2013) in a publication regarding festival tourism, where we will find a description of the influence of festival tourism in terms of its functions and dysfunctions with reference to tourism space, which is treated as one of the geographical subspaces. Cudny (2013, p. 107) quoted

Liszewski (1995, p. 94), according to whom tourism space is a "functionally distinctive part of geographical space, consisting of the natural elements of the Earth's crust (natural environment), permanent effects of human activity in this environment (economic environment), as well as the human environment in the social sense. Tourism space understood in this way is a functionally distinctive subspace of geographical and social space, and the motives to create and develop it include the needs for recreation, cognition/learning and experiencing, which are attributes of contemporary man's tourism".

The outcomes of festival tourism development were presented by Cudny (2013), who divided them into elements of tourism space distinguished by Włodarczyk (2009). Festival tourism has an influence on such elements of tourism space as natural and cultural heritage, or infrastructure. It also exerts multidimensional effects on man as a subject of tourism space. All these multidimensional impacts prove that festival tourism is one of the most important elements shaping the process of festivalisation, described earlier in this chapter:

1. Impacts on natural heritage
   a. Positive (functions): promoting natural environment protection and sustainable development,
   b. Negative (disfunctions): threats to the natural environment, caused by heavier environmental pollution.
2. Impacts on cultural heritage:
   a. Positive (functions): attracting new visitors to material heritage facilities like museums or galleries, development of the local and regional cultural heritage, enabling tourists to have contact with the cultural heritage, promotion of culture including high culture,
   b. Negative (disfunctions): possibility of losing the authenticity of the local and regional heritage.
3. Impacts on infrastructure:
   a. Positive (functions): creation of facilities for the purpose of organising festivals, development of tourist infrastructure, development of accommodation and catering infrastructure for the festival tourists,
   b. Negative (disfunctions): faster wear of some elements of infrastructure caused by heavy tourist traffic or vandalism.
4. Impacts on man as a subject of tourist space:
   a. Positive (functions): festival as a place where tourists may spend their free time in an interesting way and realise their interests and festival as an area of ideas and views exchange, as well as education. Festivals as places where interpersonal relations are started and developed and social ideas are developed. Festivals as creating jobs for the local population in the festival services sector,

b. Negative (disfunctions): eventual conflicts between festival tourists and the local inhabitants, possibility of excluding a part of the unaccepted inhabitants from the festival spaces and the possibility of the emergence of different pathologies during festivals (Cudny 2013).

4.4 Festivalisation Spaces

As it was shown in the previous subchapters, festivals significantly influence various dimensions and types of urban space. Festivals transform everyday spaces into spaces of festivalisation, which are full of new possibilities, opportunities and activities. This process referred to as festivalisation is sometimes also called eventalism of space (Rofe and Woosman 2015, p. 4). The impacts of festivals are both positive and negative, and they can be observed in the space components. This is the already described festivalisation process. On the other hand, through their influence and the fact that they occupy specific spaces, festivals also create spaces which can be defined as festivalisation spaces—they are geographical subspaces, both objective (measurable) and subjective (immeasurable or hard to measure) ones (based on classification by Kostrowiecki 1997). These subspaces are affected by important tangible and intangible impacts of festival events (Table 4.3).

The measurable spaces of festivalisation are large areas occupied by festivals, which exist in real space, i.e. they are physically tangible. Thus, they are elements of the physical space distinguished by Lisowski (2003). We can mention here buildings in which festivals take place, squares and streets, green areas, as well as other similar spaces where one can stay and which can be touched and seen in the real world. Immeasurable spaces include those elements of geographical space

Table 4.3 Typology of festivalisation spaces

Typology criterion	Types of festivalisation space
Physicality (tangibility)	Measurable Immeasurable
Construction	Open Closed
Location	Rural Urban
Period during which the space was occupied for festival purposes	Spaces of permanent festivalisation Spaces of temporary festivalisation
Accessibility	Public Private Hybrid
Seasonality	Summer Winter

which are significantly influenced by festivals. However, these elements are not physical or tangible; they are also hard or sometimes impossible to measure. Based on Lisowski (2003), this type of festivalisation space contains the following sub-spaces: ecological, cultural and social.

In the first case, we deal with economic relations occurring within the space of human life (understand here as an ecological space), including the space of tourism or event services market. Here, the market as intangible space of the exchange of services connected with event organisation is a particular festivalisation space. Although the market flows of capital and services can be measured, it requires complicated research methods, the flows are not visible right away, and they are not fully defined. Therefore, they should be regarded as spaces which are difficult to measure. The immeasurable spaces also include the cultural space distinguished in its abstract aspect containing the system of cultural signs. The festival is a space of transmitting temporal cultural signs, e.g. through the song, dance, recitation, discussion and other forms of non-material artistic expression. In this way, the intangible space of presenting and exchanging such signs between artists and the audience during the festival is also a festivalisation space. It is similar in the case of social space, encompassing elements produced by the community as a part of economic, cultural or political activity (Lisowski 2003). We also deal here with a form of immeasurable festivalisation space, including interpersonal relations.

Referring to the classification of geographical space by Thrift (2003), measurable spaces are elements of the first space including empirical constructions; to a certain degree, they are an element of the place—space. The immeasurable festivalisation spaces include elements of the second (unblocking) space and the third (image) space. In the latter case, an important immeasurable festivalisation space is the media space. We mean here presenting festivals in the electronic media, such as television or the Internet, and the traditional ones, e.g. the press. In this way, a festival influences the awareness of the users of these media, integrates them and creates the image of the host area. The media space that makes such activities possible should be regarded as an immeasurable festivalisation space.

Considering their construction, measurable (physical) spaces can be divided into closed and open ones (Table 4.3). The closed ones include all structures possessing walls and a roof, e.g. sports and performance halls, culture centres, theatres, cinemas and festival centres. An example of such closed festivalisation spaces is the Theater am Potsdamer Platz called the Berlinale Palast, opened in 1999 in Berlin. Since 2000, the building has been the main venue of the Berlin International Film Festival (Internationale Filmfestspiele Berlin), held in February. Closed spaces also include sacral buildings, e.g. churches, where religious festivals, including music events, are held (Fig. 4.1). Another closed space is that of university buildings, which often host festivals of science or those devoted to geography and adventure (Fig. 4.2).

As regards open spaces, we deal with areas devoid of construction elements like walls or the roof. We may mention here parks and other green areas, streets or squares. A good example of an urban open festivalisation space is St Mark's Square, used every year during the Carnival in Venice. An intermediary type of space is represented by stadiums, where music festivals are often organised. They

Fig. 4.1 A cathedral in Łódź during the summer festival of organ and cathedral music. *Source* Author's photograph

Fig. 4.2 Łódź Technical University facilities during the Festival of Wreck Diving. *Source* Author's photograph

have walls (stalls), and sometimes are partly roofed. An example of a partly closed festivalisation space, confined with tall bleachers, is the famous Sambadrome in Rio de Janeiro. Built in 1984 and recently renovated, it is a part of one of the city streets. Along an 800-m-long section, there are currently two large bleachers, seating over 70 000 spectators. The Sambadrome is used mainly during the carnival, during which the bleachers are taken by the audience and the jury evaluating the parading schools of samba.

As regards location, festivalisation spaces can be divided into rural and urban (Table 4.3). The former include spaces situated in areas which do not have the status of the city, where rural festivals are held (for rural festivals see among others: Long and Perdue 1990; Janiskee 1991). The latter type of festivalisation space includes spaces in cities where urban festivals are organised (for urban festivals see among others: Quinn 2005; Waitt 2008; Johansson and Kociatkiewicz 2011).

Apart from rural and urban, festivalisation spaces can be divided into permanent and temporary, according to the time of being used for the purpose of a festival or a group of festivals (Table 4.3). Permanent festivalisation spaces are areas occupied by a festival or a given group of festivals, as well as by institutions working for them on permanent basis. We may talk of such spaces in a situation when a space is "occupied" by a festival event for a larger part of the year, when it does not feature other dominant activities or activities which are at least equivalent to festivals. An example of such space is the Palais des Festivals et des Congrès in Cannes, opened in 1982. At present, it hosts many events, including festivals, throughout the year, e.g. the most famous Film Festival de Cannes held in May. Other festivals include the International Festival of Creativity, Festival de l'Art Chinois, International Games Festival and Pantiero Festival (see http://en.palaisdesfestivals.com/).

As regards temporary festivalisation space, we talk about a situation when the space is taken up by other activities, unrelated to festivals for most of the year. Then, the festival function only supplements other activities performed within a given space. Perfect examples of this type of festivalisation space are public spaces, such as city streets or squares, used for festival purposes several times a year at the most. For the remaining part of the year, they are shopping or transport spaces. An example of a temporary festivalisation space is the centre of Swedish town of Umea. Every year, the European Union holds a competition and chooses two cities, which gain a right to organise various events for the period of one year. Through these events, including numerous festivals, the cities present their cultural heritage, promote themselves and try to attract tourists. Umea was such a European Capital of Culture in 2014. In that year, the city hosted many cultural and entertainment events, including 9 festivals (see http://umea2014.se/en/project/). Some of them took place in the city centre, on a square next to the city hall. This space normally serves trading and transport purposes, while for the duration of those events, it became a temporary festivalisation space (Fig. 4.3).

As regards accessibility as a criterion of festival space classification, they are divided into public and private spaces (Table 4.3). "In common parlance, public space is associated with parks, playgrounds, or systems of open space that are obviously in the public realm" (Banerjee 2001, p. 2). As Rouba (2011, pp. 158–159)

Fig. 4.3 City square as a temporary festivalisation space—night festival events Umea 2014. *Source* Author's photograph

wrote, "according to the classical definition, public space is a non-private space, accessible for everybody; it originated in the Antiquity (the agora). Examples of public space include city squares, streets, office buildings, museums, shops, elements of natural environment, e.g. parks. Its opposite is private space, i.e. a place whose accessibility depends on our own decision, e.g. our flat or house.[1] The features of public space are security, variety, changeability, and the ability to satisfy the needs of different groups of people (Augustyn-Lendzion 2005; Sierecka-Nowakowska 2001). Public spaces also include private areas used by the public, because what makes a space public is not the form of ownership but the way it is used (Markowski 2001)."

Public spaces are places of contacts with other people. They may be contacts of low or high social density. Those which involve going out, seeing and listening to other people (hear and see contacts) are low-density contacts. On the other hand, they may involve meeting friends, neighbours and colleagues or going out with family—then, they are high-density contacts. Public spaces are also areas of making and maintaining completely new contacts. They are places where we develop social capital, thanks to observation and participation in various events. We can experience extraordinary and interesting events through meeting and establishing contacts with other people (Gehl 2011). These spaces are also an area of relaxation and fun including organisation of various events, festivals included. Carr et al. (1992, p. 48) quote a number of examples of special events organised in public spaces, such as parades of different national minorities, gay and lesbian marches and marathons.

[1]There are also intermediary spaces—half-private and half-public, e.g. staircases in blocks of flats and city yards.

It must be stressed that public spaces most frequently used for festival purposes include streets and squares in the city centre, as well as various green areas, e.g. city parks. Festivals are also often organised in hybrid spaces, e.g. within the premises of service and shopping centres (they will be described in the next paragraph). Examples of such public spaces where various festivals are held can be found all over the world. We may quote here St Patrick's Festival Parade in Dublin (Ireland), which is organised in the capital of Ireland on 17th of March, i.e. during the national holiday dedicated to the patron saint of Ireland—St Patrick. The festival in Dublin takes place on the streets and squares in the city centre, so it is an example of a festival happening in public spaces (http://www.stpatricksfestival.ie/info). A similar parade dedicated to St Patrick is organised in New York (USA), though it does not have the character of a national holiday in this country. However, it is also an example of a festival taking place in a public space, in the city centre (see http://www.nycstpatricksparade.org/about.html).

Another example of a festival organised in a public space, in the streets of a city centre, is the Birmingham's Frankfurt Christmas Market and Craft Fair. It has been organised annually, in November and December for over a decade, in one of the largest British cities. It is visited by thousands of visitors from Great Britain and abroad. Every year, about 180 stalls are put there, where you can buy original German Christmas goods (Fig. 4.4). Currently, this festival is claimed to be the largest outdoor Christmas Market in Britain. The event is accompanied by many small and large art events (http://www.birmingham.gov.uk/frankfurtmarket).

In Poland, recent years have been marked by a growing number and variety of urban festivals (see Szymańska 2007). A city where many festivals are organised every year is Łódź—the third largest city in Poland. Also there, a lot of events take place in the streets of the city centre, easily accessible for the public. We may quote here the example of the Festival of Kinetic Art of Light—Light Move Festival (see http://lightmovefestival.pl/about-the-festival/). It is a light and sound performance taking place in the evening in the space of central Łódź streets and squares, as well as a city park (a green area in the centre of Łódź, open to the public) (Fig. 4.5).

Regular or one-off festival events called "town days" are organised in many smaller towns around Łódź. They are cultural and entertainment local events which are to add splendour to the anniversary of founding a given destination. An example of such an event, organised in the public space of a park, was the Brzeziny Days in summer 2014 (Fig. 4.6) or Piotrków Trybunalski Days, organised on the streets of the Old Town (Fig. 4.7).

Particular open public spaces used temporarily for festival purposes are the spaces of rivers, water reservoirs and coastal areas. An example here is the festival of the Loire River, held in the French city of Orleans. The festival includes a parade of river boats, concerts and fireworks shows (see http://int.rendezvousenfrance.com/en/events/festival-loire-orleans). A similar event is organised on the Mississippi River in New Orleans. It is called the Downriver Mighty Mississippi River Festival (see http://www.frenchmarket.org/event/currents-the-mighty-mississippi-river-festival/). As regards festivals organised at the seaside, we may quote an event organised on the Baltic Sea in Poland, called the Baltic Sail (see http://www.zagle.com.

Fig. 4.4 Birmingham's Frankfurt Christmas Market and Craft Fair in 2014. *Source* Author's photograph

pl/wydarzenia/18-zlot-zaglowcow-i-festiwal-morski-baltic-sail-gdansk-3072014-co-nas-dzis-czeka,1_14627.html). The festivalisation space includes the river, the dock, the harbour and the river quay, as well as the area of harbour towns, temporarily used for the purpose of some festival events (e.g. ship parades or music concerts).

Nowadays, as a result of the liberalisation of the socio-economic relations in contemporary cities, we may observe a crisis of public space. Many cities do not care for its development properly, and some of its parts are privatised. In this way, some ways of using it become impossible and some of its users do not have access to it. The rules of the liberal market economy determine the shape of many contemporary cities. Therefore, the number of private spaces or privatised public spaces is growing. The latter are sometimes referred to as "hybrid spaces". They are created as a result of the cooperation between public and private investors, and managed by corporations. An example is the space of shopping centres, often including open spaces, e.g. inside squares or streets[2] (see Carr et al. 1992). They are

[2]The author of this book believes that the spaces of shopping centres should not be (as it was described in the work by Rouba 2011—quoted earlier) classified as fully public spaces.

Fig. 4.5 Festival of Kinetic Art of Light in Łódź—Light Move Festival—city centre events in 2014. *Source* Author's photograph

Fig. 4.6 Brzeziny Days in 2014. *Source* Author's photograph

Fig. 4.7 Piotrków Trybunalski Days in 2014. *Source* Author's photograph

generally commonly accessible spaces, with some exceptions: the poor, the homeless and street artists are chased away by security. These spaces are monitored, fenced, as well as often locked, e.g. at the time when the shopping centre is closed. Therefore, hybrid spaces are not fully public spaces. A perfect example of a hybrid space is a festival marketplace. It is a kind of shopping centre created in the second half of the twentieth century in the USA. It is usually located within the city centre. Apart from the shopping space, the festival marketplace includes areas accessible to the public, e.g. the central square or the street which the facility opens to. In this case, we can talk about a hybrid space, mentioned above, halfway between a private and a public one. According to Goss (1996), such a space is not, however, a fully public space but in a sense its imitation, which is actually managed by private corporations. An important feature of a festival marketplace is a well-developed culture and entertainment function. The streets or squares making up a festival marketplace host festivals of different types, which are to make the commercial offer of such a centre more attractive (see Amendola 1995; Goss 1996; White 2005; Griffin et al. 2008).

A very good example of a shopping centre of the festival marketplace type, situated in a revitalised central post-industrial area, is the Manufaktura Shopping and Service Centre in Łódź (Poland). Its important part is the centrally located square, surrounded with shopping facilities, covering over 30 000 m2. It is an example of a hybrid public–private space, which, despite being situated on the premises of a private shopping and service centre, is accessible to customers practically for 24 h, without many limitations. On the other hand, however, it is a space excluding the homeless and street performers as it is monitored around the clock. It is the venue of various types of events, festivals included, e.g. large urban festivals such as the Łódź of Four Cultures Festival (see http://www.4kultury.pl/en/

Fig. 4.8 Trotuart Festival 2014—Manufaktura Shopping and Service Centre Agora. *Source* Author's photograph

Fig. 4.9 Folk Inspirations Festival 2014—Manufaktura Shopping and Service Centre Agora. *Source* Author's photograph

festival/about-festival/), or smaller events, such as Trotuart—the festival of street theatres (see http://www.trotuart.pl/E/ETrotuArt.html) (Fig. 4.8), or Folk Inspirations—the festival of young people's art. The first of them presents street theatres and the second one—young singers, dancers, instrumentalists, artists, photographers, etc., experimenting artistically on the basis of folk culture, folklore and ethno (Fig. 4.9) (see http://www.folkoweinspiracje.pl/en/).

Another example of a hybrid space is the area of Łódź Street Food Festival. It is organised in the city centre, in Piotrkowska Street, on the premises of a former textile factory. The area is the property of a private firm, which rents its buildings and spaces for various commercial purposes. It includes shops and restaurants, and several times a year, a food festival is held there. It is important that the event is open to everybody, it is not ticketed, and you do not have to buy anything. Thus, it is an area open to the public, though during the festival it is monitored and protected by the organisers. It is not accessible for 24 h as it is closed for the night; thus, it is also an example of a hybrid festivalisation space (Fig. 4.10).

Fig. 4.10 Łódź Street Food Festival, spring 2015. *Source* Author's photograph

Another type of festivalisation space is private space, where festivals are also frequently organised. As Kirby (2008, p. 84) wrote, "Private space is maintained by capital and structured by civil law, it is necessarily different from traditional spaces (such as parks), which are directly administered by the local state, via its ordinances and ultimately, criminal statute (Frug 1999). It is clearly subject to different forms of control, which are both more subtle and more binding. Often, they are based on barriers that are both literal (a lack of pedestrian access for example) and monetary (including some costs of entry, such as parking, in the case of malls, or fees and dues in more permanent arrangements such as residential developments). Surveillance and policing are usually undertaken by private security firms, which are more restricted in their ability to use force than traditional police".

It can be seen that in the case of private space, we deal with an area which is cut from other areas of the city with clear borders. It is a fenced, protected area which is not accessible to anyone who wants to enter it; entry often requires the owner's permission or buying a ticket. In this case, we deal with, e.g., a private house as a festivalisation space. Other examples are closed facilities, such as a shopping centre or theatre.

Examples of closed, isolated, private spaces are culture clubs or festival centres, as well as cinemas or theatres where theatre or film festivals are held. They are fully monitored and protected; to attend a festival organised in such space, it is necessary to buy a ticket. For instance, Warsaw hosts Warsaw Film Festival every year in October (see http://www.wff.pl/en/). It takes place in several spaces of the city, e.g. on the premises of a large culture and entertainment centre situated in the city centre, i.e. in the Palace of Culture and Science. Some shows take place in a multiplex, which is a part of a large shopping centre situated nearby, called Golden Terraces. These spaces are typical examples of private festivalisation spaces (Figs. 4.11 and 4.12).

Fig. 4.11 Warsaw Film Festival 2014—Palace of Culture and Science. *Source* Author's photograph

Fig. 4.12 Warsaw Film Festival 2014—multiplex at the Golden Terraces shopping centre. *Source* Author's photograph

Very interesting festivalisation spaces are experience and tourism centres, situated at large car factories. Such centres are built by large car companies, e.g. Volkswagen, in order to promote the brand and attract potential customers. A good example of such a centre is Autostadt, situated at a large Volkswagen factory in a

Fig. 4.13 Autostadt (Wolfsburg) during one of the cultural events in 2014

German town of Wolfsburg. The centre, which receives about two million visitors every year, presents the technological achievements of the company, as well as its car models. However, an important component that is to make a visit to Autostadt more attractive is its well-developed cultural and entertainment function. The centre offers numerous concerts, fireworks shows or water shows. Moreover, it is an area hosting festivals, such as the Dance Festival—Moviementos. In this way, the area of Autostadt becomes a private festivalisation space. The visitors' zone is delimited, fully monitored and protected; the entry is possible after buying a ticket (see http://www.autostadt.de/en/ort/) (Fig. 4.13).

The next criterion of the festivalisation space classification is seasonality (Table 4.3). In this case, we may refer to the classification of festivals quoted earlier, e.g. into summer and winter events (see Sect. 2.3). Festivalisation spaces can also be divided into summer, winter, spring and autumn ones, as well as into all-year-round spaces of festivalisation.

Study tips:

- Explain the concept of geographical space and its main types.
- Define urban space.
- Describe the concept of festivalisation.

- Present the most important positive and negative impacts of festivals on individual elements of urban space.
- Define the features of festival tourism compared to other types of tourism.
- Define and divide festivalisation spaces.

References

Aldskogius H (1993) Festivals and meets: the place of music “Summer Sweden”. Geografiska Ann Ser B Hum Geography 75(2):55–72

Alzua A, O’Leary J, Morrison A (1998) Cultural and heritage tourism: identifying niches for international travelers. J Travel Tour Stud 9(2):2–13

Amendola G (1995) Public spaces and city regeneration. Culture and neighbourhoods, concepts and references, vol 1. Council of Europe Publishing, Strasbourg, pp 71–79

Andersson I (2014) Placing place branding: an analysis of an emerging research field in human geography. Geografisk Tidsskrift-Danish J Geogr 114(2):143–155

Arcodia C, Cohen S (2007) Environmental accreditation for event management. In: Proceedings of the 4th international event management research conference re-evaluating the city/town: events as a catalyst for change, Melbourne. http://www.business.uts.edu.au/acem/pdfs/proceedings_international_event_symposium_2007.pdf/. Accessed 18 Sep 2011

Arcodia C, Whitford M (2006) Festival attendance and the development of social capital. J. Convention Event Tour 8(2):1–18

Atkinson D (2005) Heritage. In: Atkinson D, Jackson P, Sibley D, Washbourne N (eds) Cultural geography: a critical dictionary of key concepts. I.B.Tauris, London, pp 141–152

Augustyn-Lendzion E (2005) Przestrzeń publiczna? Kilka refleksji na temat. Czasopismo Techniczne Architektura, 102(9-A):35–44

Banerjee T (2001) The future of public space. J Am Plan Assoc 67:9–23

Bjeljac Ž, Terzić A, Lović S (2015) Nematerijalno kulturno nasleđe u Srbiji kao turistički brend. Turističko poslovanje 15:53–61

Boland P (2010) Capital of culture—you must be having a laugh!’ Challenging the official rhetoric of Liverpool as the 2008 European cultural capital. Soc Cult Geogr (7):627–645

Boo S, Busser JA (2006) Impact analysis of a tourism festival on tourists destination images. Event Manag 9(4):223–237

Bourdieu P (1984) Distinction: a social critique of the judgment of taste. Harvard University Press, Cambridge, Massachusetts

Buch T, Milne S, Dickson G (2011) Multiple stakeholder perspectives on cultural events: Auckland's Pasifika festival. J Hosp Market Manag 20 (3–4):311–328

Buczkowska K (2008) Turystyka kulturowa. Przewodnik metodyczny, Eugeniusz Piasecki Physical Education Academy in Poznań, Poznań

Carlsen J (2009) The economics and evaluation of festivals an events. In: Yeoman I, Robertson M, Ali-Knight A, Drummond S, McMahon-Beattie U (eds) Festival and events management. Elsevier, Oxford, pp 249–259

Carr S, Francis M, Rivlin LG, Ston AM (1992) Public space. Cambridge University Press, Cambridge

Casanova H, Hernandez J (2011) The regeneration of public life. Scape 1:18–25

Caves RE (2000) Creative industries. Harvard University Press, Cambridge and London, Contracts between art and commerce

Cooper C, Hall CM (2008) Contemporary tourism: an international approach. Butterworth-Heinemann, Burlington

Crompton JL, McKay SL (1994) Measuring the economic impact of festivals and events: some myths, misapplications, and ethical dilemmas. Festival Manag Event Tour 2:33–43

Cudny W (2006) Festiwal Dialogu Czterech Kultur jako przykład strategicznego produktu turystycznego dla Łodzi. Turystyka i Hotelarstwo 10:117–128

Cudny W (2011) Film festivals in Łódź as a main component of urban cultural tourism. Bullet Geogr Socio–Econ Ser 15:131–141

Cudny W (2012) Introduction. In: Cudny W, Michalski T, Rouba R (eds) Tourism and the transformation of large cities in the post—communist countries of Central and Eastern Europe ŁTN. Wydawnictwo Uniwersytetu Łódzkiego, Łódź, pp 9–19

Cudny W (2013) Festival tourism—the concept, key functions and dysfunctions in the context of tourism geography studies. Geografický časopis 6(2):105–118

Cudny W (2014) Festivals as a subject for geographical research. Geografisk Tidsskrift-Danish J Geogr 114(2):132–142

Cudny W, Rouba R (2011) Theatre and multicultural festivals in Lodz as a free time management factor among the inhabitants and tourists in the post-industrial city. Acta Geographica Universitatis Comenianae 55(1):3–22

Cudny W, Gosik B, Rouba R (2011) Festiwal kultury chrześcijańskiej w Łodzi jako przykład produktu turystycznego—wydarzenie. Ekonomiczne Problemy Usług 66(648):89–100

Cudny W, Michalski T, Rouba R (eds) (2012a) Tourism and the transformation of large cities in the post—communist countries of Central and Eastern Europe. ŁTN—Wydawnictwo Uniwersytetu Łódzkiego, Łódź

Cudny W, Korec P, Rouba R (2012b) Resident's perception of festivals—the case study of Lodz. Sociológia—Slovak. Sociol Rev 44(6):704–728

Cunningham SD (2002) From cultural to creative industries: theory, industry, and policy implications, media international Australia incorporating culture and policy. Q J Media Res Resour 54–65

Dallen JT, Boyd SW (2003) Heritage tourism. Pearson Education, Harlow

Davies M (1993) Cultural tourism, history, and historic precincts. In: Hall MC, McArthur S (eds) Heritage management in New Zealand and Australia. Oxford University Press, Auckland, pp 188–196

Davies WKD (2015) Festive cities: multi-dimensional perspectives. In: Davies WKD (ed) Theme cities: solutions for urban problems. Springer, Cham, pp. 533–561

de Valck M (2007) Film festivals: from European geopolitics to global cinephilia. Amsterdam University Press, Amsterdam

Derrett R (2003) Making sense of how festivals demonstrate a community's sense of place. Event Manag 8:49–58

Dimitrić D, Bjeljac Ž (2015) Structure and motives of visitors to the event "Evenings of Tešnjar" J Geogr Inst "Jovan Cvijić" SASA, 65, 2, pp 215–227

Dinnie K (2011) City branding: theory and cases. Basingstoke, Hampshire, Houndmills [England]; Palgrave Macmillan, New York

Du Cros H, Mckercher B (2002) Cultural tourism: the partnership between tourism and cultural heritage manager. Haworth Hospitality Press, New York

Dwyer L, Mellor R, Mistillis N, Mules T (2000) A framework for assessing 'tangible' and 'intangible' impacts of events and conventions. Event Manag 6(3):175–189

Elliott A, Urry J (2010) Mobile lives. Routledge, London—New York

Flew T (2010) Toward a cultural economic geography of creative industries and urban development: introduction to the special issue on creative industries and urban development. Inform Soc (26):85–91

Florida R (2003) Cities and the creative class. City Commun 2(1):3–19

Fredline L, Derry M, Jago L (2006) Host community perceptions of the impact of events. A comparison of different event themes in urban and regional communities, CRC for sustainable tourism Pty Ltd. Gold Coast, Australia. http://www.crctourism.com.au. Accessed 1 Sept 2011

Frey B (1986) The salzburg festival: an economic point of view. J Cult Econ 10:27–44
Frey B (1994) The economics of music festival. J of Cult Econ 18:29–39
Frug G (1999) City making: building communities without walls. Princeton University Press, Princeton NJ
Gaworecki WW (2007) Turystyka. Warszawa, PWE
Gehl J (2011) Life between buildings: using public space. Islandpress, Washington
Getz D (1991) Festivals, special events, and tourism. Van Nostrand Reinhold, New York
Getz D (2008) Event tourism: definition, evolution, and research. Tour Manag 29(3):403–428
Getz D (2010) The nature and scope of festival studies. Int J Manag Res 5(1):1–47
Gnoth J (2004) Strengthening tourism SME brands, proceedings: Rencontres de St Gall, Sept. 20–23, Appenzell, Switzerland. In: Fueglistaller U, Volery T, Weber W (eds) Value creation in entrepreneurship and SMEs. KMU, Hochschule St Gallen
Goss J (1996) Disquiet on the waterfront: reflections on nostalgia and utopia in the urban archetypes of festival marketplaces. Urban Geogr. 17(3):221–247
Gotham KF (2005) Theorizing urban spectacles. Festivals, tourism and the transformation of urban space. City 9(2):225–246
Govers R, Go FM (2009) Place branding: local, virtual and physical identities, constructed, imagined and experienced. Palgrave Macmillan, Basingstoke
Griffin T, Hayllar B, Edwards D (2008) Places and people: a precinct typology. In: Hayllar B, Griffin T, Edwards D (eds) City spaces. Tourist places, Urban Tourism Precincts, Butterworth-Heinmann, Oxford, pp 39–62
Hagen J (2008) Parades, public space, and propaganda: the Nazi culture parades in Munich. Geografiska Ann Ser B Hum Geogr 90(4):349–367
Hall CM (1989) Hallmark events and the planning process. In: Syme GJ, Shaw BJ, Fenton DM, Mueller WS (eds) The planning and evaluation of hallmark events. Aldershot, Avebury, pp 20–42
Hall CM (1992) Hallmark tourist events: impacts, management and planning. Bethaven Press, London
Hauptfleisch T, Lev-Aladgem S, Martin J, Sauter W, Schoenmakers H (eds) (2007) Festivalising!: theatrical events, politics and culture. Rodopi, Amsterdam—New York
Holt M (2005) Young people and illicit drug use in Australia. Soc Res Issues Paper 3:1–4
Hughe V, Stall R, Klouri C, Barrett DC (1995) AIDS risk-taking behavior during carnival in São Paulo, Brazil. AIDS 9(1):39–44
Hughes G (1999) Urban revitalization: the use of festive time strategies. Leisure Stud 18(2):119–135
Ivanovic M (2008) Cultural tourism. Juta & Co., Cape Town, South Africa
Janiskee R (1991) Rural festivals in South Carolina. J Cult Geogr 11(2):31–43
Jensen OB (2007) Culture stories understanding cultural urban branding. Plan Theory 6:211–235
Johansson M, Kociatkiewicz J (2011) City festivals: creativity and control in staged urban experiences. Eur Urban Reg Stud 18(4):392–405
Kaczmarek J, Stasiak A, Włodarczyk B (2010) Produkt turystyczny. Pomysł, organizacja, zarządzanie. PWE, Warszawa
Kang S, Kim K, Ryan C, Park S (2014) What makes people travel to cultural heritage festival? Int J Tour Sci 14(3):70–87
Kirby A (2008) The production of private space and its implications for urban social relations. Polit Geogr 27(1):74–95
Kostrowicki A (1997) Przestrzeń—jej istota i zróżnicowanie. Rzeki 6:125–139
Kotler P, Armstrong G (2010) Principles of marketing. Upper Saddle River, N.J., Pearson Education, London
Kowalczyk A (2001) Geografia turyzmu. Wydawnictwo Naukowe PWN, Warszawa
Larson M, Wikströ M (2001) Organizing events: Managing conflicts and consensus in a political market square. Event Manag 7:51–65
Lee CK, Lee YK, Wicks BE (2004) Segmentation of festival motivation by nationality and satisfaction. Tour Manag 25(1):61–70

Lee I, Arcodia C, Lee TJ (2010) The key characteristics of multicultural festivals: a critical review of the literature. In: Rhodri, T. (ed) Events and festivals research: state of the art. Global events congress IV: festivals & events research: state of the art. Incorporating the 8th AEME events Management Educators' Forum, Leeds, U.K.13–16 July 2010, pp 1–10. http://www.eventsandfestivalsresearch.com/files/proceedings/LEE%20ARCODIA%20&%20LEE%20FINAL.pdf. Accessed 10 Sept 2011

Lerner J (2003) Acupunctura urbana. Editora Record, Rio de Janeiro

Leszczycki S (1975) Geografia jako nauka i wiedza stosowana. Państwowe Wydawnictwo Naukowe, Warszawa

Lew AA (1987) A Framework of Tourist Attraction Research. Ann Tour Res 14:553–575

Lisowski A (2003) Koncepcje przestrzeni w geografii człowieka. Uniwersytet Warszawski, Warszawa

Liszewski S (1995) Przestrzeń turystyczna. Turyzm 5:87–103

Liszewski S (2008) Formy i struktury przestrzenne wielkich skupisk miejskichw: Liszewski, S. ed. Geografia urbanistyczna, Wydawnictwo Uniwersytetu Łódzkiego, Łódź, pp. 187–294

Long P, Perdue T (1990) The economic impact of rural festivals and special events: assessing the spatial distribution of expenditures. J Travel Res 28(4):10–14

Long P, Robinson M (eds) (2004) Festivals and tourism: marketing, management and evaluation. Business Education Publishers Ltd., Sunderland

Markowski T (2001) Przestrzeń publiczna w ekonomice rozwoju miast. Biuletyn KPZK PAN, z. 194

Markwell K, Waitt G (2009) Festivals, space and sexuality: gay pride in Australia. Tour Geogr 11 (2):143–168

Marston S (1989) Public rituals and community power: St. Patrick's parades in Lowell, Massachusetts. Polit Geogr Q 8(3):1841–1874

McCarthy J (2006) Promoting image and identity in "Cultural Quarters": the case of Dundee. Local Econ 20(3):280–293

McClinchey KA (2008) Urban ethnic festivals, neighborhoods, and the multiple realities of marketing place. J Travel Tour Market (3–4):251–264

Mezias S, Pedersen S, Svejenova S, Mazza C (2008) Much ado about Nothing? Untangling the impact of European premier film festivals. Creative Encounters Working Paper, 14:2–31

Mezias S, Pedersen SJ, Kim JH, Svejenova S, Mazza C (2011) Transforming film product identities: the status effects of European premier film festivals, 1996-2005. In: Moeran B, Pedersen SJ (eds) Negotiating values in the creative industries: fairs, festivals and competitive events. Cambridge University Press, Cambridge, New York, pp 169–196

Mikkonen J, Pasanen K (2010) Economic and socio—cultural impacts of two cultural festivals of Savonlinna. In: Paper for global events congress IV: events and festival research: state of the art, Leeds. http://www.eventsandfestivalsresearch.com/files/proceedings/Mikkonen_Pasanen_FINAL.pdf. Accessed 25 Dec 2011

Miles S (2005) Understanding the cultural case: class, identity and the regeneration of Newcastle Gateshead. Sociology 39(5):1019–1028

Nuryanti W (1996) Heritage and postmodern tourism. Ann Tour Res 23:249–260

Ooi CS (2002a) Cultural tourism and tourism cultures: the business of mediating experiences in Copenhagen and Singapore. Copenhagen Business School Press, Copenhagen

Ooi CS (2002b) Poetics and politics of destination branding: Denmark. Scand J Hosp Tour 4 (2):107–128

Ooi CS (2010) Branding cities, changing societies. Creative Encounters Working Paper 45:3–17

Ooi CS, Pedersen S (2010) City branding and film festivals: Re-evaluating stakeholders' relations. Place Branding Public Diplomacy 6(4):316–332

Pasanen K, Taskinen H, Mikkonen J (2009) Impacts of cultural events in eastern Finland—development of a Finnish event evaluation tool. Scandi J Hosp Tour 9(2–3):112–129

Poria Y, Butler R, Airey D (2001) Clarifying heritage tourism: a comment. Ann Tour Res 28:1047–1049

Pratt AC (2009) Urban regeneration: from the arts "feel good" factor to the cultural economy: a case study of Hoxton, London. Urban Studies 46(5–6):1041–1061

Prentice R, Andersen V (2003) Festival as creative destination. Ann Tour Res 30(1):7–30
Quinn B (2005) Arts festivals and the city. Urban Stud 42(5–6):927–943
Quinn B (2006) Problematising "Festival Tourism": arts festivals and sustainable development in Ireland. J Sustain Tour 14(3):288–306
Quinn B (2010) Arts festivals, urban tourism and cultural policy. J Policy Res Tour Leisure Events 2(3):264–279
Reisinger Y (2009) International tourism: cultures and behavior. Butterworth—Heinemann, Oxford
Richard G, Palmer R (2010) Eventful cities: cultural management and urban revitalization. Butterworth-Heinemann, Oxford
Richards G (1996) Cultural tourism in Europe. CAB International, Wallingford, Oxon
Richards G (ed) (2007) Cultural tourism: global and local perspectives. Haworth Hospitality Press, New York
Ritchie B (1984) Assessing the impact of hallmark events: conceptual and research issues. J Travel Res 23(1):2–11
Roche M (2011) Festivalization, cosmopolitanism and European culture: on the socio-cultural significance of mega—events. In: Giorgi L, Sassatelli M, Delanty G, (eds) Festivals and the cultural public sphere. Routledge, London—New York, pp 124–141
Rofe MW, Woosnam CL (2015) Festivals as a vehicle for place promotion: cars, contestation and the creative city ethos, landscape research, article published online, doi:10.1080/01426397.2015.1078457
Rouba R (2011) Łódź Art Center jako nowa publiczna przestrzeń kultury w Łodzi. In: Jażdżewska I (ed) Przestrzeń publiczna miast. XXIV Konwersatorium Wiedzy o Mieście, Wydawnictwo Uniwersytetu Łódzkiego, Łódź, pp 157–172
Ryan C (1998) Economic impacts of small events: estimates and determinants—a New Zealand example. Tour Econ 4(4):339–352
Ryan C (2003) Recreational tourism: demand and impacts. Channel View Publications, Clevedon
Ryan C, Gu H (2010) Constructionism and culture in research: Understandings of the fourth Buddhist Festival, Wutaishan, China. Tour Manag 31(2):167–178
Saayman M, Saayman A (2004) Economic impact of cultural events. South Afr J Econ Manag Sci 7(4):629–641
Schlenker K, Foley C, Getz D (2010) ENCORE festival and event evaluation kit: review and redevelopment, CRC for Sustainable Tourism Pty. Ltd., Gold Coast, Australia. http://www.crctourism.com.au. Accessed 1 Sept 2011
Sharma V, Joshi BD (2010) Assessment of noise pollution during Deepwali festival in a small township of Haridwar City of Uttarakhand, India. Environmentalist 30:216–218
Sierecka-Nowakowska B (2001) Dziedzictwo przyrodniczo—kulturowe jako podstawa rozwoju przestrzeni publicznej miasta na przykładzie Łodzi. Biuletyn KPZK PAN, z. 194
Siwek T (2011) Percepce geografického prostoru. Česká geografická společnost, Praha
Smith MK (2009) Issues in Cultural Tourism Studies. Routledge, New York, Milton Park, Abingdon, Oxon
Stebbins R (1996) Cultural tourism as serious leisure. Ann Tour Res 23:948–950
Szwichtenberg A (ed) (2000) Podstawy turystyki. Politechnika Koszalińska, Koszalin
Szymańska D (2007) Urbanizacja na świecie. Wydawnictwo Naukowe PWN, Warszawa
Tetłak K (2013) The taxpayer as the unofficial sponsor of the London 2012 olympic games. Int Sports Law J 13(1):97–103
Tetłak K (2014) Sochi 2014 Olympic tax legislation. Eur Taxation 54(4):165–167
Thomas NJ, Hawkins H, Harvey DC (2010) The geographies of the creative industries: scale, clusters and connectivity. Geography 95(1):14–21
Thrift N (2003) Space: the fundamental stuff of geography. In: Rice S, Valentine G (eds) Key concepts in geography Holloway SL. Sage, London, pp 95–108
Tighe AJ (1986) The arts/tourism partnership. J Travel Res 24(3):2–5
Van Elderen PL (1997) Suddenly one summer: a sociological portrait of the Joensuu festival. Joensuu University Press, Joensuu

Waitt G (2008) Urban festivals: geographies of hype, helplessness and hope. Geogr Compass 2 (2):513–537

Wall G, Mathieson A (2006) Tourism: change, impacts, and opportunities. Pearson Education Limited, Harlow

White C (2005) Festival marketplaces. In: Caves RW (ed) Encyclopedia of the city. Routledge, Oxon, Abingdon, pp 179–180

Włodarczyk B (2009) Przestrzeń turystyczna—istota, koncepcje, determinanty rozwoju. Wydawnictwo Uniwersytetu Łódzkiego, Łódź

Zukin P (2010) Naked city: the death and life of authentic urban places. Oxford University Press, Oxford—New York

http://www.britannica.com/EBchecked/topic/557313/space.Online Encyclopedia Britannica, entry: space. Accessed 03 March 2015

http://www.trotuart.pl/E/ETrotuArt.html webpage devoted to the Trotu Art festival. Accessed 03 March 2015

http://www.4kultury.pl/en/festival/about-festival/ webpage devoted to the Lodz 4 Cultures Festival. Accessed 03 March 2015

http://www.folkoweinspiracje.pl/en/ webpage devoted to the International Festival of Youth Creativity FOLK INSPIRATION. Accessed 03 March 2015

http://www.stpatricksfestival.ie/info webpage devoted to the St. Patrick's festival. Accessed 05 March 2015

http://www.nycstpatricksparade.org/about.html webpage devoted to the St. Patrick's festival in New York. Accessed 05 March 2015

http://www.birmingham.gov.uk/frankfurtmarket webpage devoted to the Frankfurtmarket in Birmingham. Accessed 05 March 2015

http://lightmovefestival.pl/about-the-festival/ webpage devoted to the Lightmovefestival. Accessed 05 March 2015

http://umea2014.se/en/project/ webpage devoted to the European Capital of Culture—Umea 2014. Accessed 05 March 2015

http://en.palaisdesfestivals.com/ webpage presenting the Palais des Festivals in Cannes. Accessed 05 March 2015

http://www.wff.pl/en/ webpage devoted to the Warsaw Film Festival. Accessed 05 March 2015

http://www.autostadt.de/en/ort/ webpage presenting the Autostadt centre in Wolfsburg. Accessed 07 March 2015

http://int.rendezvousenfrance.com/en/events/festival-loire-orleans webpage devoted to the Festival de Loire in Orléans. Accessed 07 March 2015

http://www.frenchmarket.org/event/currents-the-mighty-mississippi-river-festival/ webpage devoted to the-mighty-mississippi-river-festival. Accessed 07 March 2015

http://www.zagle.com.pl/wydarzenia/18-zlot-zaglowcow-i-festiwal-morski-baltic-sail-gdansk-3072014-co-nas-dzis-czeka,1_14627.html webpage devoted to Festiwal Morski Baltic Sail Gdańsk. Accessed 07 March 2015

Chapter 5
The Role of Festivals in the Development of a Post-socialist City—Łódź Case Study

Abstract In this chapter, the city of Łódź and the festivals held there are presented in detail. It is then a particular case study showing a review of urban festivals and their influence on space. Łódź is a city in central Poland, inhabited by about 700 000 people. It is an example of a post-industrial and post-socialist centre undergoing a deep transformation. In 1945–1989, Poland was ruled by a communist system, imposed by the Soviet Union. Due to the "Solidarity" social movement, in 1989, peaceful reforms were introduced, which resulted in the implementation of market economy and democracy. That was how the process of transformation, i.e., the transition from the communist system to capitalist economy and a democratic state began in Poland. The transformation had a number of both positive and negative effects, which also affected Polish cities. After 1989, the changes in Łódź included a decline of the textile industry, the development of economy based on services, as well as significant spatial transformations. Łódź as an old industrial city based on traditional textile production was particularly strongly affected by the process of deindustrialisation (after 1989) and the resulting socio-economic problems. On the other hand, the free market and the introduction of democracy triggered social activity, also in the field of culture and festival organisation. There appeared many institutions and associations concerned with culture, art and entertainment. City authorities noticed that the activity known in the West as the cultural economy was successfully used for the purpose of restructuring post-industrial cities. Therefore, they began to support the development of culture in Łódź. This strategy, based on cultural and entertainment events, became one of the ways of developing the city. This led to the growth of the festival function in the city. This chapter presents a detailed description of Łódź as a city—its location, natural conditions and history. It is a broad background for the description of the major festival events held in Łódź, presented further in the chapter. This part of the book is based on own author's research and other sources. The reader can find here description of 48 urban festivals and the evaluation of their influence on the urban space of Łódź.

Keywords Post-socialist city · Culture · Festivals · Festivals in post-socialist city · Łódź · Poland · System transformation

W. Cudny, *Festivalisation of Urban Spaces*, Springer Geography,
DOI 10.1007/978-3-319-31997-1_5

5.1 The Location of Łódź

The city of Łódź covers about 293 km^2 and is situated in the centre of Poland (Fig. 5.1), in Łódź Province (*województwo*). About 47 km north of Łódź, there is the geometrical centre of Poland.

According to Liszewski (2009, p. 14), considering its administrative borders, the city lies in the area which belongs to Central Poland Lowlands (Wieczorkowska and Gozdzik 2002). This lowland location is a positive city development factor. Łódź is situated in the border area between two huge geographical units, i.e. the Masovian Lowland and the Wielkopolska (Greater Poland) Lowland. From the south, this zone borders on an upland area. The area of Łódź does not exceed the lowland altitude, which is 300 m above sea level.

Łódź is very conveniently located; the distance from the city to the furthest borders of the country does not exceed 350–400 km (Liszewski 2009, p. 20). Łódź is situated in the middle of the country, not far from the capital of Poland and other major urban agglomerations. The distance to Warsaw is about 130 km, to Poznań—about 218 km, Wrocław—216 km, Krakow—260 km and Gdańsk—about 331 km. These distances may be covered by car in 1.5–3.5 h. Thus, the accessibility of Łódź from the main Polish agglomerations is very good.

As it can be seen from the above, Łódź is located in a zone where the main Polish roads cross. North of Łódź there is the largest junction of motorways in Poland. It is where the two main motorways—A-1 and A-2—meet. Due to that, Łódź became a transport node connecting the routes running from the north to the south and from the east to the west of Europe. The areas close to the motorways are of great interest to entrepreneurs. It is where large logistic centres are built. The

Fig. 5.1 The location of Łódź. *Source* Author's compilation

Fig. 5.2 Transport location of Łódź (2015). *Source* Author's compilation based on (http://www.gddkia.gov.pl/)

roads are also very important to industry development. Road transport is often valued higher than rail transport. The good condition of the road infrastructure system is a significant factor of locating economic activity in Łódź and Łódź Province (see Fig. 5.2).

Another important aspect of the transport location of Łódź is related to the railroad system and air transport. In 2009, although Łódź had rail connections with all large cities in Poland, they were all outdated and not modernised. Only the railway line between Łódź and Warsaw has undergone the first stage of modernisation so far. Łódź has also an international airport of regional importance, called The Władysław Reymont Airport (Fig. 5.3). This airport provides several connections with Western Europe as well as charter flights to few well-known tourist destinations.

5.2 History and Contemporary Times

An important issue that needs to be discussed when presenting Łódź as a city is its history. In the historical development of Łódź, we may distinguish seven periods of different origins. They differ as regards the city-creative factors as well as the population and spatial development of the city. The first of the periods was the period of agricultural Łódź, followed by the industrial period (Table 5.1). The peak of agricultural Łódź development (the settlement was granted on ca. 1332) was the period of 1550–1650. The town, inhabited by 650–800 people, became a local

Fig. 5.3 Władysław Reymont Airport in Łódź. *Source* Author's photograph

Table 5.1 Periods in the historical development of Łódź

Years	Historical–economic period
1332–1820	Agricultural Łódź
1821–1913	Development of capitalist industrial Łódź including the peak capitalist period, starting from 1866
1914–1918	First World War—economic and social crisis
1918–1939	Independent Second Republic of Poland, revival based on textile industry
1939–1945	Second World War—economic and social crisis
1945–1989	Real socialism period (Polish People's Republic), strengthening the structure of a traditional industrial city
Since 1989	Third Republic of Poland—an intensive socio-economic transformation

Source Author's compilation, Koter (1969)

centre catering for the needs of the nearby villages located among forests, as regards craftwork and trade. However, this settlement had only local significance and never develops higher functions (Koter 1969, p. 28).

The next and very important stage in the development of Łódź was the industrial period, which began in the early nineteenth century. The history of modern Łódź is connected with the development of the textile industry in the Polish territory of that time (Koter 1969). As a result of an unfavourable political situation, Poland ceased to exist as a state at the end of the eighteenth century due to the third partition of the country by Austria, Prussia and Russia in 1795. On that territory where Łódź is now located, Russians created in 1815 a state called the Kingdom of Poland, which was dependent on the Russian Empire. Despite the fact that its politics and economy were strictly connected with the Russian ones, it had some autonomy, e.g. as regarded economic issues. The Kingdom of Poland was ruled by a government consisting of Poles (see Koter 2009). At the same time, Western Europe was undergoing the industrial revolution and implemented its inventions. The lack of industry was seen in the Kingdom of Poland as the main reason of its economic difficulties. Therefore, an intensive industrialisation programme was introduced on its territory.

The government of the Kingdom of Poland devised a programme of creating industrial towns. In order to choose the most suitable areas, various parts of the country were inspected. It was decided that the area around agricultural Łódź was suitable to create there a few textile industry settlements (since 1821) (Koter 1969). Thanks to industry, Łódź became a centre attracting immigrants from the whole region as well as abroad. The city turned to a so-called promised land, a place where in favourable circumstances one could make a fortune manufacturing and selling textile goods. As a consequence, the number of population increased rapidly, especially in 1820–1914. Over that time, the number of population in Łódź grew from 767 to 477 990 people (over 623 times) (Dzieciuchowicz 2009a, p. 114). As regards the nationality structure of Łódź population, as a result of the inflow of immigrants from Poland and other regions of Europe (german countries and Russia), it was highly diversified and changeable. It is estimated that in 1831 the population of the city consisted of about 74.1 % of Germans, 17.4 % of Poles and about 8.5 % of Jews. In 1857, there were 43 % of Poles, 41 % of Germans and about 15 % of Jews. Other nationalities, including Russians made up about 1 % of the inhabitants of Łódź. In 1897, Łódź was inhabited by 46.4 % of Polish people, 29.4 % of Jews, 21.4 % of Germans and about 2.4 % of Russians, as well as 0.4 % of the representatives of other nationalities (Dzieciuchowicz 2009a, p. 122).

The First World War brought substantial material losses in the city due to the German occupation of the area where Łódź was situated. Germans occupied the area previously annexed by Russians because they were winning the war on the eastern front. The policy of the German occupants was extremely brutal; they requisitioned and stole property. As a result, they also confiscated textile raw materials (e.g. cotton), ready-made products as well as stole factory machinery (Table 5.1).

As a consequence of the defeat of Germany and Austria-Hungary, the fall of Russia after the Revolution, as well as the liberty movement, Poland regained independence in 1918. The period of the Second Republic of Poland began. At that time, the textile industry was revived; it started to produce goods, mainly for the domestic market. It was a clear change in the direction of the sale of products, which before 1914 were mostly exported to Russia. It was mainly the result of the Soviet Union's hostile attitude to Poland, which limited economic contacts between the two countries. After the Polish–Bolshevik war (1919–1921), trade contacts with the east were practically suspended. This made it difficult for Łódź industry to function normally because the city lost its sales and commodity market, as most cotton processed in Łódź came from the East. The interwar period was a time of stabilisation and socio-economic revival, but it was also a time of slowing down due to the problems with export to the Soviet Union (Jewtuchowicz and Suliborski 2009b).

On 1 September 1939, Poland was invaded by the German army, and on 17 September 1939, the Russian army entered its territory. This is how the Second

World War started. Łódź was taken by Germans on 9 September 1939 and renamed as Litzmannstadt (Nartonowicz-Kot 2009, pp. 190–191). It was the beginning of the worst period in the history of Łódź. During the Nazi occupation, Germans committed countless crimes on the population and destroyed the city economically. Between 1939 and 1945, the number of the population of Łódź within its prewar boundaries decreased from 680 000 to 250 000. The losses reached 63.2 % of the number of Łódź inhabitants. According to Puś (1987), about 279 000 people were murdered, including about 170 000 Jews living in the city (concentrated in the Łódź Jewish Ghetto established by Germans in 1940). Further losses concerned the people who were displaced from Łódź, escaped from Hitler's terror or were taken to concentration camps or for compulsory labour in Germany (Dzieciuchowicz 2009b, p. 276–277).

Russian soldiers entered Łódź on 19 January 1945 and that is how the city was liberated from the German occupation. 1945 was also the moment when the Soviet Union introduced the undemocratic communist system, also known as real socialism, in Poland. From that moment until 1989, Poland was an incapacitated country, politically and economically dependent on Soviet Russia. The economically inefficient system of centrally planned economy was introduced, in which decisions concerning the type and location of economic activity were taken by the state. All larger enterprises were nationalised; private property functioned only in small craftsmanship, services and agriculture (Nartonowicz-Kot 2009, p. 199). In 1945–1989, Łódź based its development on the traditional textile industry, the produced goods including natural and artificial fibres and fabrics. Related industries developed as well, producing, e.g. chemicals, clothes or textile machinery, exported first of all to the Soviet Union and to other communist countries. Regarding the structure of inhabitants of the city in the post-war period, Łódź became a mono-national city inhabited mostly by Poles. The Jewish population was murdered by Germans occupants during the war and the population with German roots fled from Łódź when the Soviet Army was approaching in 1945 (Dzieciuchowicz 2009b, p. 279).

In 1989, in Poland, we witnessed a change of the political, economic and social system, which involved transforming the undemocratic socialist system into a democratic one, combined with the introduction of the market economy. After 1989, Łódź became a typical post-socialist and post-industrial city (for a description of post-socialist cities, see Matlovic et al. 2001, Ondoš and Korec 2008; Marcińczak and Sagan 2011; Cudny 2012). A post-socialist city is a centre which for several decades functioned in the undemocratic system of real socialism. It was subject to management processes consistent with the rules of centrally planned economy (Liszewski 2001). After the fall of this system, such city entered the phase of deep, multidimensional transformations (Sýkora 2009), these transformations could also be seen in Łódź. They encompassed the institutional sphere (e.g. democratisation, privatisation, liberalisation of prices and trade), as well as social (e.g. economic restructurisation, socio-economic liberalisation, social polarisation, cultural changes) and spatial (e.g. commercialisation of the city centre, revitalisation, suburbanisation) ones (see Sykora and Bouzarovski 2012).

Łódź was one of the cities most strongly affected by the process of transformation. The city was then a centre based on traditional industry. Its characteristic feature was the very unfavourable domination of a single branch (the textile industry), supplemented with other related branches. The level of services was low; thus, there was not really any type of activity other than textile production in Łódź. The inflow of cheap textile products from Asia, no export to Russia, overemployment and wrong management of state enterprises, as well as the total lack of the government's help at that time, were the reasons why the economic base of the city was shaken. The textile industry collapsed, the city started to suffer from mass unemployment and poverty. As a result of the growing negative migration balance and negative population growth rate, the number of Łódź population dropped below 750 000 in 2009. It was about 100 000 inhabitants fewer than in 1985. The result of the demographic crisis was also the intensive ageing of the city population (Cudny 2012).

It was only in the late 1990s and the next decade that it was possible to observe some improvement of the economic situation, privatisation and an inflow of foreign investors. Thanks to that, industry was restructured and the service sector started to flourish. As a result of the economic changes, the unemployment in the city decreased significantly (from about 72 000 in 1992 to about 22 000 in 2008) (Cudny 2012, p. 13). After Poland joined the European Union in 2004, EU funds helped to develop. Unfortunately, the strong crisis, affecting the city in the 1990s and the loss of the economic base still makes Łódź a city displaying much worse social and economic indexes than the remaining Polish urban centres (e.g. Poznań, Wrocław, Krakow or Warsaw) (see Cudny 2012; Cudny and Rouba 2012b).

At present, it seems that Łódź has come out of the worst phase of the crisis and is in a state of some economic balance. However, its demographic–social indexes are still very low. It also seems that the level of earnings, which is lower than the national mean and the mean for other large cities, is still a big problem in the city. In order to solve these problems, the city tries to attract new investors, focuses on developing the innovation sector, as well as tries to gain a better image. In the 1990s, there were attempts to use the culture-led revitalisation strategy for the development of Łódź. It included the development of a variety of festivals (see Cudny 2006). They were to be a driving force behind the development of tourism in the city. Similar strategies were implemented in West European countries (see Evans 2003; Garcia 2004). Łódź (with different outcomes) followed the example of Manchester, where after the decline of the textile industry, cultural and artistic events started to be organised, which were to intensify tourism and improve the perception of the city.

The strategy of developing cultural and entertainment events implemented in Łódź after 1989 was connected with the attempts made by the city authorities to promote Łódź as a tourist city. According to many research (see Young and Kaczmarek 2008; Cudny and Rouba 2012b), Poles perceived Łódź in a very negative way and the citizens of other countries practically did not know anything about the city. During the communist period, the authorities did not see tourism as a significant element of urban economy. Industrial tourism was not popular in Poland at all, while cultural tourism was associated with other cities, such as Warsaw or Krakow. As a result, Łódź was not perceived as a tourist city, but rather as a colourless, backward and poor centre of traditional industry. After 1989, this odium of a negative perception of Łódź remained in the Poles' awareness and it is very difficult to change it.

5.3 Major Festivals

Urban Culture in Łódź
Łódź used to be and sometimes still is perceived as a non-cultural city, a kind of desert without any significance on the culture map of Poland. However, analysing the historical development of culture in the city as well as the evolution of education and science, we can notice how untrue this statement is. Where do such negative opinions about the third most populated city in Poland come from? On the one hand, they come from its historical past, and on the other—from the unfavourable image of Łódź, strengthened by its history as the city of labourers. As shown by Kamiński (1962, pp. 267–268), Łódź was a huge city which developed as a result of the huge migration from the surrounding villages in the nineteenth century. A consequence of that was the long-lasting predominance of simple and uneducated industrial workers of rural origins in the city community. For instance, according to a national census, in 1931 Łódź occupied the penultimate position among Polish cities inhabited by over 100 000 people as regarded the number of white collar workers. In 1950, the percentage of illiterate people in Łódź was by 50 % higher than the average illiteracy rate among the urban population in Poland. As regarded the development of cultural institutions, the situation of Łódź was worse than of other large cities, such as Krakow, Warsaw or Poznań. It was due to the short history of the city, which had been unable to create so many institutions of this kind over only 130 years. Other cities simply had a longer history; they were already large urban centres in the Middle Ages and the Renaissance so they managed to create many more cultural or scientific institutions, as well as form the intelligentsia social group. Łódź did not become a big city until the middle of the nineteenth century and had much less time to develop such institutions (Kamiński 1962).

Secondly, during the Second World War, Łódź suffered tragic population losses, though its buildings survived without much damage. In the case of other cities, the damage to the urban tissue was enormous. In Warsaw, for example, over 80 % of the urban essence was destroyed. After the war, such cities underwent intensive revitalisation, contrary to Łódź. Moreover, the development of post-war Łódź was based on outdated textile industry, which brought little income and did not develop new solutions. This also caused a particular socio-economic stagnation in the period of communist Poland. All those factors made Łódź a centre which was negatively perceived in Poland and practically unrecognised abroad (see Young and Kaczmarek 2008). This does not mean, however, that the city was devoid of cultural institutions, cultural and entertainment events or scientific institutions and the intelligentsia. Even at the time of partitions (before 1918), the cultural and entertainment activity was present in Łódź, e.g. connected with the numerous ethnic groups inhabiting the city. In the 1850s, Łódź had many amateur theatre groups, mainly Polish and German. In 1866, the city gained its first permanent theatre, followed by other ones in the following decades. The first musical society was founded in 1846; in 1887 the first concert hall was built. In the interwar period, Łódź music conservatory was founded. Numerous newspapers published in the city represented Poles, Jews and Germans living there. The arts flourished, including sculpture and painting at which Jewish artists excelled. All those institutions and organisations became an element shaping the multinational cultural heritage of industrial Łódź (Koter et al. 2005). In 1896, the first open-air cinema show in Łódź was organised, and in 1899, the first stationary cinema was open in Piotrkowska Street. In 1808, in Old Town the first elementary school was founded, followed by another one opened in 1829. The first secondary school was started in 1847 and the first public library was opened in 1917 (Jewtuchowicz and Suliborski 2009a, pp. 152–158).

In the nineteenth and the early twentieth century, an important role in the development of culture and education was played by Łódź factory owners. They donated substantial sums to the building, organisation and maintaining cultural and scientific institutions. They built schools, and supported music groups and cultural enterprises financially. We should mention here the greatest Łódź industrialists, such as Karol Scheibler, Henryk Grohman or Israel Poznański, who were often called the creators of industrial Łódź.

Following the First World War (after 1918), culture and education in Łódź entered the phase of metropolitan development. Cultural and educational institutions of a higher rank developed, many of them receiving organisational and financial support from the authorities. Elementary education developed, e.g. in 1938, there were 199 elementary schools, 122 of which were financed by the city authorities. In the late 1930s, there were 32 secondary schools in Łódź, most of them private. In the interwar period, an equivalent of a university was founded in the city, called the Free Polish University. Furthermore, a number of scientific associations were established, e.g. historical, geographical or literary, and the network of libraries was visibly growing. In 1918–1939, there were 34 cinemas in Łódź. There were also a considerable number of newspapers, including those of national range. In mid-1920s, a vibrant City Theatres Group was created. Music societies were developing as well. In the interwar period, four museums and three art galleries were opened in Łódź. In 1939, the Second World War broke out and Germans entered Łódź. They immediately started to impose the repression policy as regarded culture and education. Educational functions as well as culture and entertainment were forbidden to Poles and Jews, so Polish and Jewish newspapers, theatres, cultural societies (e.g. concerned with theatre or music) and art galleries were closed down. Museum exhibits were confiscated; newspapers and schools were closed down. In this way, the cruel period of occupation left its sinister mark on Łódź culture, education and science (Jewtuchowicz and Suliborski 2009b, pp. 323–333).

The year 1945 marked the beginning of the post-war revival of culture, education and science in Łódź. It was the year when the University of Łódź, the Technical University of Łódź and Medical Academy were founded. Other similar institutions were established after the war included the present Grażyna and Kiejstut Bacewicz Musical Academy or Władysław Strzemiński Academy of Fine Arts. Many research institutes were founded, including a branch of the Polish Academy of Science. As a result, the city became a major academic centre, educating students also in the field of art and high culture. For instance, in the 1959/1960 academic year, Łódź was the fifth city in Poland with the largest number of students (Dylik and Dylik 1962).

An additional important element of the development of culture and science in Łódź was founding the Higher Film School in 1948. Today, it is called the Polish National Film, Television and Theatre School (Państwowa Wyższa Szkoła Filmowa, Telewizyjna i Teatralna—PWSFTviT). Moreover, due to the war damages in Warsaw, it was decided to open the first post-war film studio in Łódź. In 1945, the Feature Films Studio was launched, which remained one of the largest film studios in Poland until 1989. In 1949, the Educational Film Studio (1949) and the Se-Ma-For Studio (1956) were founded. The former one produced nature films, documentaries and popular science films, while the latter was the biggest Polish producer of puppet and film animation (Cudny 2011).

After 1945, Łódź gained a number of other culture-related investments and institutions, such as:

- Arthur Rubinstein Łódź Philharmonic, established in 1949 (its predecessor was Łódź Symphonic Orchestra founded in 1915).
- the Grand Theatre (1967)—one of the most important and the largest opera and ballet stages in Poland.
- 32 cinemas in 1985.
- after 1945 there appeared many cultural centres, art ensembles, e.g. musical ones, which were subsidised from the budgets of huge state textile enterprises (Cudny and Rouba 2012b).

The 1990s were a period of the most intensive transformations following the fall of communism. During the first stage of the transformation, Łódź culture experienced deep crisis. Many of the institutions established in 1945–1989 collapsed, mainly due to financial reasons. It concerned, e.g. culture centres or ensembles financed by large enterprises, which went bankrupt after 1989. Cudny and Rouba (2012b) show that due to the decline of the main textile factories, many culture clubs (there were 350 of them in the 1970s) were closed down. The number of cinemas decreased considerably, which was also caused by the popularisation of video, and later DVD and satellite TV technology. Unfortunately, as a consequence of the film sector entering the market, as well as the domestic competition with the film production centre in Warsaw developing in the 1990s, the Łódź Film Centre weakened and other film institutions, such as Se-Ma-For producing animated films, lost their significance (see Cudny and Rouba 2012b; Cudny and Ogórek 2014).

On the other hand, the democratisation of social life, introducing the freedom of speech and assembly, as well as the growing role of local government, started to have a positive influence on the urban culture. In 2012, in Łódź, there were 15 theatres and a concert hall (in a newly built facility), 9 cinemas, 81 galleries and 16 culture centres. In 2012, the renovation of the Grand Theatre—the main Łódź theatre—began. The number of museums increased after 1989, and some of the old institutions were moved to new addresses. An important event was also the opening of the Open-Air Museum of Timber Construction in 2008, presenting Łódź buildings from the early nineteenth century (Cudny and Rouba 2012b).

A part of the system transformation in post-socialist cities was the development of festivalisation. The changes which took place in centres such as Łódź in fact reflected the changes in the cities of Western Europe in the second half of the twentieth century. They were transformations which were a part of the transition from the fordist to post-Fordist economy. In post-socialist cities, they naturally took place later (for political reasons), were more intense and took less time than similar changes in Western countries (see Ondoš and Korec 2008). The transformations were connected with the transition from the economy based on traditional industry to one based on services, knowledge and creativity, as well as a modern and flexible production sector. One of the elements of such transformation was the development of the cultural and entertainment activity and using it for changing the cities that were once based on traditional industry. The culture and entertainment sector was to become an element of development which was to replace the declining industry, at least in part. Such strategies were called culture-led regeneration strategies (their detailed description can be found in the previous chapters). A part of them was the development of festivals, which were to generate tourism, improve the image of the city and offer a way of spending free time to its inhabitants (see Cudny and Stanik 2013).

A similar strategy, based on culture and festivals in particular, started to be implemented in Łódź already in the 1990s. At that time, the urban festivals sector was well developed; they were co-financed by the city authorities. They were organised by various associations and non-public institutions, though there was also a large group of festivals organised by self-governing institutions, e.g. museums or theatres (see Cudny 2006). It seems that this kind of strategy brought some results, though not as far reaching as it had been expected. Łódź became a destination, which to some extent is perceived as a festival city, especially as regards film festivals, as these events are undoubtedly connected with the cultural heritage of Film Łódź. We must remember that after 1945, the city became a place where the most outstanding Polish film-makers were educated and where important Polish pieces of film art were produced (Domański 2008).

On the other hand, many festivals held in the city are small events, which do not generate considerable tourist traffic. They do not contribute much to the promotion of the city in Poland and abroad. Many of the festivals organised in the city are visited by the inhabitants of Łódź mostly; thus, they rather fulfil the needs and expectations of local community than tourists. Considering that fact we must admit that the development strategy based on the creation of festival sector was not fully successful. However on the other hand, many of main aims of festival development are connected with local community. They may fulfil the needs connected with contact with culture, development of social capital, entertainment and fun. One of the biggest problems for culture in the city is conflicts in the cultural sector. In 2010, a serious conflict antagonised the city authorities and the organisers of the largest Łódź festivals. It was a conflict of political and financial nature, involving the organisers of the Dialogue of Four Cultures Łódź Festival and the organisers of the Camerimage Film Festival. As a result, Camerimage—a festival well known in Poland and abroad—was moved to the city of Bydgoszcz. The conflict was surrounded with an aura of scandal, which strongly undermined the reputation of Łódź festivals and worsened the image of the city (Cudny 2011).

In the next years, a new city development strategy was created. It was decided that the future of Łódź should involve creative development, based on science, creativity and implementation of new solutions. The city slogan that was coined was "Creative Łódź". Moreover, it was decided to restructure the expenditures on cultural and entertainment events. The city was to finance mainly the flagship events related to the creative sector. One of those was a fashion event—Łódź Fashion Week, often described as the festival of fashion. Some related events are also supported, such as the competition for the best fashion designer, entitled Golden Thread. An important festival which receives a lot of support is the Design Festival, dedicated to the art of industrial design. These events are regarded as significant from the point of view of promotion and they are subsidised from the city promotion fund. Besides, the money for culture development (also the parts of culture oriented towards satisfying the city inhabitants' needs) is allocated to a number of smaller events (see Klima and Rosińska-Bukowska 2012).

Łódź Festival Events

One of the aims of this part of the chapter is to briefly describe the festivals held in Łódź in recent years. Based on his own research and field observation, as well as a review of publications concerning Łódź events, the author conducted an analysis of these festivals. Their number, structure and seasonality, as well as the results of studies concerning the organisational and financial structure and the development perspectives of Łódź events will be presented here. The research was conducted among the organisers, and it also concerned the perception of festivals among the inhabitants of Łódź.

It should be stressed that Łódź is not an exception among large Polish cities. Other cities with over 500 000 inhabitants, such as Warsaw, Wrocław or Krakow, also organise many large festivals every year. Stanislawska (2007 cited in Cudny 2014, p. 646) reports that in 2004, about 100 large festivals were organised in Krakow, about 80 in Warsaw and about 70 in Wrocław. According to Pawlusiński (2007, p. 222), Krakow hosted the total of about 1100 cultural events, such as exhibitions, festivals and concerts. This number includes several dozen festivals, well fitting into the realities of this city. Krakow is the former capital of Poland, famous for numerous historical monuments, the oldest Polish university, as well as a well-developed sector of culture and entertainment and multicultural historical heritage. As a result, the number of annual events, including festivals, held there is among the greatest in Poland.

Generally speaking, Polish cities have a very rich history and culture. As a result, they have been hosting music or film festivals for many years. However, after the fall of communism in 1989, there appeared new, better conditions for the development of festival activity in Poland. First of all, social and economic freedom was granted. Many institutions, associations and foundations organising festivals were established. People could pursue their passions and interests in this way. Local self-governments were empowered, which enabled them to implement their own policy of supporting events. New companies were set up, which provide professional services during events, including festivals. Interest in this type of events increased among the society and they started to be perceived as a possible source of income and a tool for creating a positive image of cities. Hence, the demand for festival events increased (see Jędrysiak 2008; Buczkowska 2009).

The first detailed studies of Łódź festivals were carried out after 2000. According to the study published by Stanisławska (2007), in 2004, 60 regular festivals took place in Łódź. The research was based on a survey conducted at various institutions and on the analysis of the Internet data. The majority of Łódź events were devoted to art and science (21 %), theatre (18 %) and music (18 %). A smaller percentage concerned regional and local culture (10 %), poetry (8 %), film and photography (8 %), as well as nature, history and other themes (17 % together). The main events mentioned in that publication (this research was based n data from 2004) included the Dialogue of Four Cultures Festival, dedicated to the multicultural heritage of the city (ca. 100 000 spectators in first edition, 400 performers), the Camerimage—a cinematographers' festival (ca. 45 000 spectators, 2500 guests), as well as the Explorers Festival, devoted to exploration, tourism and adventure (ca. 10 000 spectators, 30 guests), Łódź Biennale art festival (ca. 8000 visitors, 93 artists), The 60th Anniversary of the Liquidation of the Litzmannstadt Ghetto (ca. 5500 guests), the regional Jarmark Wojewódzki (Province Fairs) (ca. 5000 visitors), and the A. Tansman International Festival and Competition of Music Individualities (ca. 3000 spectators, 40 performers).

Based on the survey conducted by Stanisławska (2007) among Łódź inhabitants and the participants of seven chosen festivals, the influence of the events was evaluated. The results show that Łódź events were attended mainly by the inhabitants of the city and the Łódź region (Łódź Province). In 2004, the audience coming from that area made up 84 % of all visitors. The remaining ones came mainly from other regions of Poland, e.g. Masovia (3 %), Kuyavia-Pomerania (2 %), Pomerania (2 %), Lower Silesia (2 %) and other areas (2 %). The structure of the guests invited to take part in the events (artists, performers) was different. Here, depending on the event, there were many guests from abroad (mainly from Europe and the USA). As shown by the research above, Łódź festivals are mainly intended for spectators who live in Łódź and the Łódź region (usually coming for 1 day). In 2004, they did not generate much tourism, particularly from abroad.

Over the period of 2010–2011, a number of research works concerning festival events were created at the Institute of Tourism and Economic Development, the Tomaszów Mazowiecki branch of the University of Łódź. Their authors were the Institute employees and their initiator and co-author was dr Waldemar Cudny. One of those publications was an outcome of international cooperation with the geographers from the Comenius University in Bratislava in Slovakia. As a result of a series of studies, many scientific articles appeared in Polish and Slovakian journals (see Cudny 2011; Cudny and Rouba 2011a, b, 2012a, c; Cudny and Stanik 2013; Cudny and Ogórek 2014; Cudny et al. 2011, 2012). There also appeared many BA dissertations supervised by the employees of the Institute.

The material used to write this part of the chapter was obtained from the research presented above. Some of it was updated and supplemented with new information and photographs. One of the basic methods applied in order to obtain information necessary to describe these festivals was the query. It involved collecting information about individual festivals at the festival offices, the City Hall, as well as on the Internet. The query concerned the most important information about festivals, i.e. their history, organisational structure, financing, the programmes of festival events, as well as the number and structure of the visitors.

The other important method was the questionnaire method, used in the study of the perception of Łódź festivals. In 2011, 1245 questionnaires were distributed among the inhabitants of Łódź, concerning their participation in festivals and how they perceived the events held in the city (see Cudny et al. 2012). The results of those studies will also be briefly presented in this chapter.

Moreover, a detailed questionnaire survey was realised in Łódź, which regarded the segmentation and motivations of the visitors of selected Łódź festivals (see Cudny and Stanik 2013; Cudny and Ogórek 2014). Several film festivals were analysed with regard to who and why attended them. The case studies of these festivals will be briefly presented as well.

It was decided by the author to review Łódź festivals organised in recent years. The review is based on the research conducted by the author in 2009–2011 and presented in a number of publications (Cudny 2011; Cudny and Rouba 2011a, b, 2012a, c; Cudny and Stanik 2013; Cudny and Ogórek 2014; Cudny et al. 2011). Overall, 48 events of the festival type were identified; they were events organised regularly in 2009–2011 in the area of Łódź. As regards their history, a part of them are rooted in the communist period, but the majority of them were created after the fall of communism in 1989. According to Cudny and Ogórek (2014, pp. 45–46), 80 % of music festivals and 55 % of theatre festivals, as well as 77 % of festivals devoted to other arts were created after 1989.

The Łódź festivals described in this subchapter were divided into several types, on the basis of a classification presented by Cudny (2011) and Cudny and Rouba (2012c). The festivals were divided according to their rank into mega-events (high rank and large scale, present in international media, influencing the economy of the city, region, or even country), distinctive events (high rank, various scales, commonly recognisable and identified with the spirit of the city or region), large-scale events (large scale, lower rank, quite popular, having large audiences), local events (low rank and small scale of events intended mainly for the visitors from the region). Most of the events described above (Table 5.2) are art festivals, i.e. events devoted to film (12 events—25 %), theatre (11 events—23 %), music (10 events—21 %) or graphic design (4 events—8 %). Other festival themes included

Table 5.2 Łódź festivals organised in 2009–2011

No.	Topic	Festival name	Classification by rank
1	Theatre	Międzynarodowy Festiwal Sztuki Ulicznej—Trotuart (International Festival of Street Art—Trotuart)	Distinctive
2	Theatre	Spotkania Teatrów Miast Partnerskich (Meetings of Partner Cities Theatres)	Distinctive
3	Theatre	Nowa Klasyka Europy—Międzynarodowy Festiwal Teatralny Klasyki Światowej. (New European Classics—International Theatre Festival of World Classics)	Distinctive
4	Theatre	Łódzkie Spotkania Baletowe—Międzynarodowy Festiwal Sztuki Baletowej (Łódź Ballet Meetings—International Festival of Ballet Art)	Distinctive
5	Theatre	Międzynarodowy Festiwal Solistów lalkarzy. (International Festival of Solo Puppeteers)	Distinctive
6	Theatre	Ogólnopolski Festiwal Sztuk Przyjemnych i Nieprzyjemnych (Polish Festival of Pleasant and Unpleasant Arts)	Distinctive
7	Theatre	Międzynarodowe Biennale—Spotkania Teatralne Terapia i Teatr (International Biennale—Therapy and Theatre Meetings)	Distinctive
8	Theatre	Festiwal Szkół Teatralnych (Theatre Schools Festival)	Distinctive
9	Theatre	Łódzkie Spotkania Teatralne (Łódź Theatre Meetings)	Distinctive
10	Theatre	Ogólnopolski Przegląd Teatrów Dziecięcych Dziatwa (Polish Review of Children's Theatres—"Dziatwa")	Distinctivee
11	Theatre	ŁóPTA—Łódzki przegląd teatrów amatorskich. (Łódź review of amateur theatres)	Local
12	Photography	Międzynarodowy Festiwal Fotografii -Fotofestiwal (International Festival of Photography)	Distinctive
13	Graphics	Międzynarodowy Festiwal Komiksu i Gier (International Festival of the Comic Book and Games)	Distinctive
14	Various arts	Łódź Biennale	Dostinctive
15	Artistic fabrics	Międzynarodowe Triennale Tkaniny (International Fabric Triennale)	Distinctive
16	Graphics	Międzynarodowe Triennale Małe Formy Grafiki (International Triennale: Small Forms of Graphic Design)	Distinctive
17	Various arts	Letni Festiwal Sztuki (Summer Art Festival)	Local
18	Various arts	Festiwal Sztuk Cygańskie Impresje (Art Festival: Gypsy Impressions)	Local

(continued)

Table 5.2 (continued)

No.	Topic	Festival name	Classification by rank
19	Folk art	Międzynarodowe Warsztaty Folklorystyczne (International Folklore Workshops)	Distinctive
20	Other	Kongres Iluzjonistów (Magicians' Congress)	Distinctive
21	Film	Off jak gorąco (Off it's hot!)	Distinctive
22	Film	Festiwal Twórców Powiększenie (The "Large Format" Film Makers Festival)	Distinctive
23	Film	The ReAnimation Festival	Distinctive
24	Film	Festiwal Kina Letniego Filmobranie (The "Film-picking" Summer Cinema Festival)	Local
25	Film	The Cinergia Festival—the European Cinema Forum	Large scale
26	Film	Ogólnopolski Festival Filmów o Rodzinie FOR (Polish Festival of Films about the Family)	Distinctive
27	Film	Festiwal Kino Tańca (Dance Cinema Festival)	Local
28	Film	The ArtMediaTravel Festival	Local
29	Film	Włodzimierz Puchalski Nature Film Festival	Distinctive
30	Film	Festiwal Mediów Człowiek w Zagrożeniu (The "Man in Danger" Media Festival)	Distinctive
31	Film	Mediaschool Międzynarodowy Festiwal Szkół Filmowych i Telewizyjnych (Mediaschool—the International Festival of Film and Television Schools)	Distinctive
32	Film	Plus Camerimage—the International Film Festival (editions 2000–2009)	Distinctive
33	Music	Międzynarodowy Festiwal i Konkurs Indywidualności Muzycznych im. Tansmana (Tansman International Festival and Competition of Music Individualities)	Distinctive
34	Music	Międzynarodowy Konkurs im. Karola Szymanowskiego (Karol Szymanowski International Competition)	Distinctive
35	Music	Wędrowny Festiwal Filharmonii Łódzkiej KOLORY POLSKI (The Wandering Festival of Łódź Philharmonics THE COLOURS OF POLAND)	Large scale
36	Music	Ogólnopolski Studencki Przegląd Piosenki Turystycznej YAPA (Polish Student Review of Tourist Song YAPA)	Distinctive
37	Music	Łódzkie Spotkania z Piosenką Żeglarską Kubryk (Łódź Meetings with Sailors' Song—*Kubryk*)	Local
38	Music	Cantio Łódźiensis (Łódź Choral Festival)	Local
39	Music	Festiwal Muzyki Filmowej (Film Music Festival)	Distinctive
40	Music	SOUNDEDIT Festival	Distinctive

(continued)

Table 5.2 (continued)

No.	Topic	Festival name	Classification by rank
41	Music	Vena Music Festival	Large scale
42	Music	Międzynarodowy Festiwal Kultury Hiszpańskojęzycznej Viva Flamenco (International Festival of Spanish Culture Viva Flamenco)	Local
43	Religion	Festiwal kultury chrześcijańskiej (Festival of Christian Culture)	Distinctive
44	Fashion	Łódź Fashion Week	Distinctive
45	Design	Łódź Design Festival	Distinctive
46	Adventure	Explorers Festival	Distinctive
47	Adventure	Wreck Diving Festival	Distinctive
48	Various arts	Łódź Czterech Kultur (Łódź of Four Cultures Festival)	Distinctive

Source Author's compilation based on Cudny (2011), Cudny and Rouba (2011a, b), (2012a, c), Cudny and Stanik (2013), Cudny and Ogórek (2014), Cudny et al. (2011)

photography, artistic design (comic books), artistic fabrics, religion, fashion design, folk art, adventure and tourism (11 festivals—23 %). As regards classification according to rank and scale, 75 % (36 festivals) of the festivals were distinctive events, 19 % (9 festivals)—large-scale events and 6 % (3 festivals) were local events. Mega-events did not appear in the studied group of festivals (Table 5.2). Distinctive festivals were usually important events for persons who were professionally involved in theatre or music. In the case of distinctive and large-scale events, the majority of spectators were the inhabitants of the city or region. As a result, these festivals were not a major tourist asset and did not attract large numbers of tourists.

As regards seasonality, most festivals in Łódź are organised from spring to autumn, and less frequently during the holiday season (July–August) and in winter months, from December to February. Such temporal distribution of festivals results from the climatic conditions. The best weather is in May to June, so open-air events can be organised then as well. In the summer holiday months, more people leave the city and there is a danger of small attendance, so the organisers are not willing to plan large festival events then. On the other hand, this period is not completely devoid of them. In the autumn period (September–November), the weather is still quite good and people have already returned from holidays, so the number of festivals held in the city grows again (see Cudny and Rouba 2011b; 2012c). These conclusions have also been confirmed in other publications concerning cultural events in the city of Łódź (see Adamus and Paluch 2012).

As regards the spaces in which Łódź festivals are held, they are mainly closed spaces, such as theatres, culture clubs, concert halls, museums, cinemas, the Atlas Arena performance hall, Łódź Art Centre. Most of these facilities can be found in the centre of Łódź, i.e. in an area with the dominant service function (including the cultural and entertainment function), and with the best developed transport system in Łódź. A smaller number of events are organised in the open air, e.g. in Piotrkowska Street or in the Manufaktura central square (see Adamus and Paluch 2012).

As regards the organisers of the festive events, they were mainly public entities, which included theatres, culture clubs, museums and Łódź Philharmonics. Festivals were financed first of all from public funds and by private sponsors. The money came mostly from the city (city budget) and the government (the budget of the Ministry of Culture and National Heritage). Such a situation was typical of theatre and music festivals (see Cudny and Rouba 2011b, 2012c). The organisers of other art festivals, e.g. those dedicated to photography, folklore or graphics (including comic books art), include city institutions (e.g. the Department of Culture at Łódź City Council, City Art Gallery) and non-governmental bodies (e.g. Łódź Art Centre, CONTUR Artists' Society). These events are financed from public donations and by private sponsors (Cudny and Rouba 2011b).

The situation of film festivals (between 2009—2011) was slightly different—some of them were also organised by public institutions (e.g. the Museum of Cinematography or State Higher School of Film, Television and Theatre), but the majority of them were held by non-government organisations. A significant generator of events was the private cinema presenting independent repertoire (Charlie Cinema) and the TUMULT foundation, organising the famous Camerimage festival (see Cudny 2011). Camerimage, however, is an example of problems that may arise between organisers and city authorities. For several years, it was the most important festival event in Łódź, but as a result of a conflict concerning financial and to some extend (local) political issues, the festival was moved to another city in 2010, in an atmosphere of scandal (Cudny 2011).

An interesting example of a Łódź film event is the Mediaschool festival, organised by the Film School since 1994. Its aim is to review the films made by the students of the leading film schools all over the world. The festival is a small event (several hundred visitors), attracting mainly people involved in film production (e.g. the students of Łódź Film School as well as other film and theatre schools in Poland). The organisers invite many guests from abroad. The survey conducted in 2011 showed that 46 % of the respondents were visitors coming from outside Łódź. About 27 % declared that they had arrived especially for the festival. Those respondents can be treated as typical festival tourists. The most important motivations to attend the festival included interest in culture and art (e.g. experiencing culture and film art) as well as making new social contacts, mainly among people connected with film production (see Cudny and Ogórek 2014).

Another example of a film festival is the Man in Danger Media Festival, first organised in 1990 by the Museum of Cinematography. The festival is devoted to documentary films concerning serious social problems. It is a regional event, which was shown in the survey, but it is important for the heritage of Film Łódź. About 80 % of the visitors questioned for the purposes of the publication by Cudny and Stanik (2013) were people involved in the film industry, such as producers, students, actors, directors or cinematographers. Thanks to the festival, they become familiar with new trends in documentary film production; they also have contact with foreign film-makers. The event also has a pedagogical dimension, because it is visited by many young people, who are shown films about serious problems in human life. In the case of the Man in Danger Media Festival, the main motivations to attend it were those related to culture and art (to experience culture and film art), event attractiveness (to see new interesting things during the festival), originality, novelty (to gain new experiences) and to social contacts (to meet people with similar interests) (Cudny and Stanik 2013).

The problems concerning the Camerimage festival, mentioned before, were not exceptional in Łódź. At the same time, similar problems occurred with another event, called The Dialogue of Four Cultures Łódź Festival. It was organised annually since 2002 by the Association for The Dialogue of Cultures, called "Łódź—the Land of the Future". The Association was based on the idea of the multicultural tradition of Łódź; its aim was to promote the city and support cultural initiatives enhancing the dialogue between different nations. In this case, it was a dialogue between the ethnic groups which were building the city in the nineteenth century. As it was mentioned in the earlier part of the chapter, the most important ethnic groups in Łódź at that time were Poles, Jews, Germans and Russians. The first edition of the festival took place at 29 different venues, its budget was about 2.5 million Euros and it was visited by about 100 000 people (see Cudny 2006, 2007). The event was an example of a multicultural festival, i.e. one which presents the culture of different nations inhabiting a given area. On the other hand, it was also an example of a restorative festival (see Duvignaud 1976), i.e. an event based on the culture of nations which at a given moment do not live in the area of Łódź any more.

In the years that followed, the event was a permanent and important element of the festival calendar of the city. The next editions were not so big and did not attract so many visitors. The event experienced some financial problems, yet, it was one of the main festivals in Łódź. Unfortunately, in 2010, a conflict broke out due to organisational and financial reasons, between the festival organisers and the city authorities. As a result, the Dialogue of Four Cultures Łódź Festival was not supported by the city any longer and stopped being organised. It was replaced by a similar event, organised by a different entity, called Łódź of Four Cultures. It seems, however, that the conflict worsened the already negative image of Łódź as a cultural city at that time. It started to be perceived as an uncertain partner for cultural institutions. The conflicts mentioned above were broadly discussed in the media and spoilt the positive image of Łódź as a city of festivals, created with so much effort after 1989. At the time of the conflict, Łódź was aspiring to receive the status of the European Capital of Culture for 2016, but it was rejected, among others due to the problems described above.

Another important Łódź festival is Łódź Fashion Week (Fig. 5.4). It may be argued whether it is more a trade-commercial event or a festival. On the one hand, Łódź Fashion Week may be treated as a business event, belonging to the MICE sector, while on the other hand it may well be treated as a fashion festival because it is not a purely business event, like, e.g. motor vehicles fairs. It contains a strong cultural component (fashion is a part of human culture) and features non-commercial shows. Moreover, it is supplemented by accompanying events, e.g.

Fig. 5.4 Events during Łódź Fashion Week. *Source* Author's photograph

a festival of films about fashion. Łódź Fashion Week displays other features typical of festivals as well, e.g. it is a meeting place and a space where we can pursue our interests. Currently (2015), Łódź Fashion Week is regarded as one of the main events in the city and supported from the city promotion budget. The city authorities treat it as a leading event in the strategy of promoting the city as a centre of creative industries related to fashion design.

A similar role is played today by the Design Festival, which is a well-known event (also abroad), dedicated to industrial design (Fig. 5.5). This festival is also subsidised from the city budget allocated to the promotion of Łódź. By supporting these events, the city realises the development strategy based on creative industries and innovativeness (the Creative Łódź strategy). These events are a slightly stronger generator of tourist traffic than the remaining art-related events. In 2011, Łódź Fashion Week was attended by over 20 000 visitors. Based on a survey conducted during the event, the percentage of tourists visiting it can be estimated at about 75 % (Cudny and Rouba 2012a).

Fig. 5.5 Events during Łódź Design Festival. *Source* Author's photograph

Important and interesting events are also festivals dedicated to tourism, adventure and exploration, such as the Explorers Festival and Wreck Diving Festival. The former is a large event (ca. 8000 visitors), oriented mostly towards the youth from Łódź and the region. The latter is a festival for the lovers of diving among ship wrecks. It is a small event, attracting tourists from different regions of Poland and guests from abroad (see Cudny and Rouba 2011a).

The International Festival of Comic Books and Games, organised in Łódź since the early 1990s has considerable potential. It is the largest event of this type in Poland, visited by several thousand participants. It is quite similar to the well-known Comic-Con International held annually in San Diego USA. The Festival of Comic Books and Games includes exhibitions, meetings with the most famous comic books artists; stands are put up where the visitors can buy comic books and games. There are also computer games competitions organised during the main event (Fig. 5.6). It must be stressed that the formula of this particular festival refers to creativity, as well as high technology. Comic books, graphics and computer games are examples of creative industries. Taking the growing number of

Fig. 5.6 The international festival of comic books and games at the Atlas Arena—the sports and performance hall in Łódź. *Source* Author's photograph

games fans into consideration, as well as the fact that the value of games market currently exceeds the value of the global film market, supporting and developing this festival may bring measurable benefits to the city.

The Evaluation of Festivals by Łódź Inhabitants and Organisers
Important publication that has to be taken under consideration in analysis conducted for the purposes of this book is a large work (already mentioned before) concerning the perception of urban festivals by the inhabitants of Łódź. Based on 1245 questionnaires distributed in 2011 in the city centre, the inhabitants' opinions about the Łódź events were studied. The results of these studies show that the inhabitants' attitude to festivals is generally positive. A great majority of the respondents believed that festivals are an interesting form of spending free time, that they have a

positive influence on the image of Łódź. The respondents remembered, however, that in 2010 a few serious conflicts grew around famous festivals. One of them (Camerimage) was moved to Bydgoszcz, while the other one (The Dialogue of Four Cultures Festival) was transformed into another event (Łódź of Four Cultures Festival). The problems were widely commented in the regional and national media and depreciated the image of the city, which did not escape the respondents' attention (Cudny et al. 2012).

As regards attendance, over half of the respondents visited festivals, mainly large and renowned art events. They also often participated in student events, such as the Festival of Student's Song—Yapa, which could result from the large number of students among the respondents. The most frequently mentioned events were two festivals which are not organised in Łódź anymore, i.e. the Dialogue of Four Cultures Festival (renamed as Łódź of Four Cultures) and Camerimage (moved to another city). It points to the fact that they had well-deserved reputation, high attendance and losing them was a great loss for Łódź.

In 2012, the author of this book supervised a study concerning the evaluation of festivals by their organisers (see Skibińska 2012). It was directed to the group of 30 festivals held in Łódź in 2012. The persons representing the institutions which organised those events (e.g. the directors of festival offices, museums, theatres) were asked to fill in a questionnaire. Representatives of the following 14 (out of 30 primally chosen) festivals agreed to take part in the survey: International Festival of Comic Books and Games, University of Łódź Juvenalia, Łódź of Four Cultures Festival, Łódź Ballet Meetings, "Dziatwa"—Polish Review of Childrens' Theatres, Theatre Schools Festival, Small Forms of Graphics International Triennale, Festival of Solo Puppeteers, International Festival of Street Art "Trotuart", Explorers Festival, "Man in Danger" Media Festival, MEDIASCHOOL—International Festival of Film Television and Theatre Schools, International Triennale—Therapy and Theatre Meetings, the Wandering Festival of the Łódź Philharmonic "The Colours of Poland".

The events listed above were mostly art festivals. The majority of them were organised in May (17 %), September (17 %), April (13 %) and November (9 %). Most of them (72 %) were annual events, the remaining ones were organised every two or three years. The venues were mostly public facilities, e.g. museums, cinemas, theatres and galleries, as well as public spaces, such as parks or streets. As many as 13 out of the studied festivals (93 %) were organised by different types of public institutions, e.g. the Łódź Culture Club, Mark Edelman Dialogue Centre, the Grand Theatre in Łódź, City Art Gallery, city culture centres. As regards the number of visitors, the answers ranged from several hundred to about 15 000. Asked about the percentage of visitors from outside Łódź, the organisers quoted 10–20 %.

As regards the answers which concerned financing, the respondents claimed that over 50 % of the funds allocated to the organisation of the festivals (even up to 90 % in some cases) came from public institutions. The biggest subsidies came from the city authorities, which was confirmed by 48 % of the respondents, then from central authorities (28 % responses), province authorities (14 % responses) and other public institutions (10 % of responses). The support by private entities ranged from several to 50 % of the money received for the organisation of the studied festivals. The largest sums were acquired from individual sponsors (57 % of the responses). Other sources of private financing were mentioned more seldom and included enterprises (21 %), foundations (7 %) and other (15 %). In the respondents' opinion (93 %), the festivals which they represented could not function without public support.

A slightly smaller percentage of the respondents (86 %) said that they could not function without the support of private sponsors. 57 % claimed that the financial support they received for the festivals was insufficient. They also claimed that insufficient financial means limit the development of festivals, e.g. by shortening the programme and decreasing its attractiveness, or creating smaller possibilities to invite interesting artists. They said that financial shortages were one of the main obstacles to the development of the events they organised. The respondents were also asked whether they received any other kind of support. They most often mentioned medial and material support, and least often—organisational support.

As regarded questions concerning the chances for and threats to the development of festivals in the future, all respondents said that Łódź was a good city for festivals to develop. Among the chances for development, they pointed to increasing the financial support for the events (which would make it possible to expand the programme, promote the event and attract more famous guests), and developing the transport infrastructure (building the new centrally located Łódź Fabryczna railway station, developing Władysław Reymont Airport). They also mentioned transplanting their events to other cities—creating so-called festival replicas. As for the threats, they most often mentioned the decreasing subsidies and insufficient number of facilities and lack of large festival centre where the events could be held.

The respondents were also asked how they ranked their events. In this case, 57 % said it was a medium-sized event, organised for the local community and tourists. 29 % ranked their event as a large festival, organised mainly for tourists and the inhabitants, and 14 % said their event was a small, local festival. The next question referred to the rank of the festival; the respondents could choose whether the festivals they organised were:

- an event at the stage of being introduced on the market,
- a very well-known product, a kind of star among Łódź festivals, but continually developing,
- a well-known event of a long tradition, presently going through a crisis, with chances for further development,
- an event of a long tradition, but facing uncertain future due to the present crisis.

The majority of respondents (64 %) chose the second variant, i.e. a festival which is a well-known product, a star, but still developing. The second most frequently chosen answer was the third variant, i.e. a festival of a long tradition, with temporary problems and chances for further development. The first variant, i.e. a local festival, was chosen least often, and the last answer was not chosen even once.

The next group of questions concerned the influence exerted by the events on the local community (as a whole), on individual inhabitants of Łódź, on the economy and the city in general. The respondents evaluated the individual kinds of impacts, ascribing them levels of significance, from insignificant through hardly significant, moderately significant, significant to highly significant. As regards the influence on the community, the highly significant impacts included the improvement of the image of the local community, help in preserving local culture, creating the sense of pride among the local community with reference to their city and its culture. According to event organisers, the most significant impact on individual inhabitants was providing them with interesting opportunities to spend their free time, enabling them to have positive relationships with tourists and increasing their cognitive possibilities. As regards the economic impact, the respondents did not say that it was significant as festivals do not generate a large number of jobs. They also often pointed to the rather unimportant role of festivals in the development of local firms and to their insignificance as regards the development of local infrastructure. As for the influence of festivals on the city, the respondents thought it was very important that events promoted the city outside and created its positive image. They also pointed to the important role of events in generating tourist traffic.

5.4 The Impact of Łódź Festivals on the Post-socialist Urban Space

Łódź is a very good example of a post-socialist city, so it is a perfect illustration of the changes occurring in cities of this type after the fall of communism in 1989. Łódź as a huge industrial centre developed in the nineteenth century which means at the time of the early capitalist economy (Koter 1969). However, the fact that after 1945 it functioned in a communist system imposed on Poland for almost 40 years, justifies calling Łódź a post-socialist city (see definition of the post-socialist city formulated by Liszewski 2001).

A post-socialist city typically undergoes a number of multidimensional transformations occurring in its area very rapidly, during a system change from communism (real socialism) into capitalist economy. That particular transformation concerns all basic urban spheres: institutional, social, functional and spatial (Matlovic et al. 2001; Sýkora and Bouzarovski 2012). It is in fact quite similar to the changes that were taking place in the second half of the twentieth century in capitalist cities, e.g. in Western Europe, where we could observe a transition from the fordist to the post-Fordist economy. The latter system was based on flexible production, development of innovativeness and modern industries, higher significance of services, including cultural and entertainment ones (Ondoš and Korec 2008). Such changes did not occur in Poland after the Second World War, because they were blocked by the authorities, for whom the most important thing was to maintain the leading role of the traditional industry and a total control over the society. As a result, processes which were natural in developed economies, such as industry restructurisation, the development of innovativeness, an increase in economic liberties, a growing demand for services and the development of this sector, did not occur in Poland.

The system transformation began in Poland in 1989, with the Round Table talks. As a result of the agreement between the representatives of the "Solidarity" opposition movement and the representatives of the communist party and the army, it was decided that political and economic reforms would be gradually introduced in Poland. The next years brought rapid development of freedom movements and the weakening of the communist power. The rules of liberal market economy and democracy were introduced. How did these processes influence the festivalisation of urban space? At the beginning, this influence was rather unfavourable, due to a serious crisis of culture in Polish cities in the early 1990s. This process was described in the previous part of the chapter, with reference to Łódź. Due to the economic crisis of the first years of the transformation, culture in Poland was financed less generously. There were no private sponsors who would support large festivals and the state enterprises which had sponsored many cultural initiatives before were declining.

On the other hand, the next years of the transformation brought gradual economic growth and an increased interest in culture, e.g. on the part of the local authorities. Their competences grew as a result of the democratisation of life and the administrative reform, which gave more rights but also delegated more responsibilities to self-governments. Many city authorities, including those in Łódź, noticed the decreasing role of industry, which shook the economic base of many Polish cities (see Słodczyk 2001). This problem was particularly significant in the case of Łódź, which before 1989 had been a city with the predominant traditional industrial function and underdeveloped services. The situation of Łódź was much worse than that of other Polish cities, where the structure of industry was more varied and services were better developed (see among others: Young and Kaczmarek 2008; Marcińczak and Sagan 2011; Cudny 2012).

Due to the deep crisis in the city, in the 1990s the local authorities started to look for a sector that might replace the declining textile industry. An obvious choice was the service sector, which started to develop due to the decreasing role of industry, the slowly growing purchasing potential of the population, or finally the consumption models adopted from the West. The authorities decided to focus on the development of the cultural and entertainment function. Their decision was the result of two factors on the one hand, the city authorities noticed that many post-industrial cities in Western Europe had implemented culture- or event-led regeneration strategies, which were a part of the cultural economy sector (see Ray and Sayer 1999; Gay and Pryke 2002; Scott 2010; Cudny and Stanik 2013). Thus, similar strategies were introduced in Łódź. Their aim was to develop the culture sector, including festivals. They were actively supported by the city, organisationally and financially. Moreover, city institutions (e.g. theatres, galleries or museums) got involved in their organisation (see Cudny 2006).

The other important factor increasing people's interest in festivals was the democratisation of social life. After several decades of the communist rule, in 1989, democracy started to be introduced in Poland, which concerned first of all the governing system, but not only. It was the time when personal and institutional freedom was restored. Personal freedom allowed people to express their opinions without constraints, follow their interests, travel, etc. Institutional freedom meant that it was possible to establish free institutions, independent of the government or any political party. It concerned both public institutions, e.g. courts, prosecutor's office, civil service institutions, territorial self-government institutions, and non-government organisations. As a result of the transformation, people were granted freedom of association, assembly, etc. There appeared numerous institutions which supported the development of culture, including non-governmental organisations, such as foundations or associations. They started to play a growing role in Łódź as regarded organising festivals. People wanted to satisfy their cognitive needs, have contact with culture, i.e. to fulfil their self-actualisation needs. The second half of the 1990s brought an improvement of the economic situation in Poland, so the people had more financial means to take advantage of events like festivals. They started to seek new experiences, enriching their minds and souls. It was the beginning of forming the experience society and the experience economy in Poland. To put it in a nutshell, a number of socio-economic and cultural factors (described in Sect. 2.2) which earlier had contributed to the festival boom in other regions of the world, appeared in Poland (including Łódź) as well.

Those factors gave a direct impulse to the development of the festivalisation process in Łódź. Apart from what has already been said, they can be divided into urban (internal) and external festivalisation factors. The internal ones were related to the new socio-economic situation—the crisis of industry and the growing role of services, including the cultural and entertainment ones. Those factors also included stronger involvement of the city authorities as well as the institutions, associations and private persons supporting festivals. Non-governmental institutions often consisted of culture and art enthusiasts, whose aim was to develop and popularise them. On the other hand, there were a number of external factors, such as the change of the global political and economic situation. At the turn of the 1980s and 1990s, the global political thaw came and the majority of communist regimes, including the Polish one, collapsed. That made it possible to change the situation on the local level. Moreover, due to the introduction of democracy and market economy, the ideas which had developed and functioned earlier in Western Europe appeared in Poland as well, including Łódź. The inflow of those ideas (e.g. cultural and experience economy, experience societies, culture—led regeneration) contributed to the process of festivalisation in Łódź.

As it was presented in Sect. 4.2, festivalisation is a complicated process of the development of festivals, which exert a significant influence on various elements of the objective and subjective urban space (see Table 4.1). In Łódź, festivalisation understood in this way took place after the fall of communism in 1989. This does not mean, of course, that culture had not developed or that festivals had not been organised in the city before. It was proved in the previous part of this chapter that culture in Łódź had been developing since the nineteenth century. Also in the period of real socialism (1945–1989), the cultural and entertainment function was increasing. However, a visible increase in the number and variety of festivals was observed after 1989.

Łódź festivals influence different elements of urban space. It is possible to demonstrate their impact on the objective (measurable) space, i.e. physical space (see Table 4.1). The city does not possess a festival centre (as of 2015) which would have changed the physical space of the city. On the other hand, many changes take place in this space during the largest festivals organised in public spaces. Temporary stages are built, where artists perform, as well as temporary stalls where you can, e.g., see or buy handicraft items. Art installations are constructed, occupying urban space. Some changes take place also in another objective space—the ecological space. As shown in Sect. 4.2, festivals have a considerable influence on many components of the ecological space. It takes the form of pollution generated by festival tourists, due to a larger consumption of water, increased emission of fumes produced by the cars which are used by tourists for transportation, an increased production of waste or larger consumption of electricity. In addition, they cause damage to the flora of the green areas where the festivals are held (parks, city forests, etc.). Regrettably, studies assessing this kind of influence on the natural environment of the city are not conducted in Łódź, unlike for instance in Australia. However, it can be estimated that with tourist traffic of about 10–20 % in the case of Łódź festivals, we may talk about some impacts on the natural environment.

It must be stressed that one of the most important impacts or functions of festivals include those referring to the types of urban subjective space (immeasurable or hard to measure), first of all the unblocking space containing socio-economic relationships, socio-economic space, as well as cultural and image space (see Table 4.1). Based on the research presented in the previous part of this chapter, it can be concluded that effects of this kind are considerable. They include both, positive impacts or functions and negative impacts or dysfunctions (for the description of festival functions and dysfunctions, see Chap. 4).

In the case of the social elements of space, we may list a number of positive impacts. They include the possibility of pursuing interests and taking part in a unique experience guaranteed by festivals. As the events held in Łódź attract mainly the city inhabitants and a smaller number of tourists, also the social impacts will concern first of all the inhabitants of Łódź. Łódź festivals represent a wide range of themes (see Table 5.2). They are art events (music, theatre, film, graphics, photography, artistic fabrics, several arts combined, etc.), events associated with exploration and adventure, or religion. As a result, the visitors may pursue various interests.

Festivals present the works of renowned artists and promote young, less famous ones. They are also places where young artists may take part in workshops (e.g. during film festivals) and make professional contacts (the festival being a point of passage). Many art-related events include competitions for the best creation, e.g. a graphic work or film, where the winners receive substantial money prizes. As regards artists of this kind, the prize is awarded not only to honour their work, but also to support financially their cultural-artistic activity. The functions described above belong to the group of impacts exerted on the social and cultural spaces. What is more, festivals enrich the culture of Łódź and make it possible to cultivate its multidimensionality and multinational historical heritage (e.g. Łódź of Four Cultures Festival). This is an important advantage of Łódź festivals. Thanks to them, high culture may be promoted (e.g. through music, theatre or ballet festivals) and the knowledge about culture and art can be popularised. Finally, a number of festivals have an ethical and didactic side to them. For instance, in the case of the film event entitled "Man in Danger", the viewers may watch documentaries devoted to human tragedies and serious social problems. As the festival is attended by quite a number of young people, it has a particularly significant didactic and educational dimension. It shows human problems and sensitises the viewers to the suffering of another person.

From the economic point of view (generating profits from tourism, the development of firms and a bigger number of jobs), Łódź festivals do not play a particularly important role. Obviously, at the time when most festivals are organised, i.e. in spring and autumn, we can observe a certain inflow of festival tourists to Łódź, but it is not significant or does not last long enough to seriously influence urban economy. As it was mentioned earlier, festival visitors from outside Łódź make up average about 10–20 %. Only some of them come from outside the region and stay in Łódź overnight—they may be treated as festival tourists. One of the most important festival events, well-known abroad and generating large tourist traffic was Camerimage, dedicated to the art of cinematography. It was an event broadly covered by the Polish media. Its organisation in Łódź was in a way connected with the idea and history of Łódź film industry. The cluster of film institutions in Łódź was named Film Łódź; the city was even referred to as HollyŁódź. However, the loss of this festival and moving it to another city spoiled the reputation of Łódź as a city of festivals. The rather small economic role of events is also confirmed by the results of the study conducted among a group of festival organisers, who described the role of events in the economic development of the city as not very significant.

The question is whether this situation must continue and what can be done to change it. It seems that focusing on the leading events (e.g. Photofestival, Design Festival or Fashion Week) in recent years has been a good move on the part of the city authorities. They are events connected with innovativeness, new ideas and design. On the other hand, they are also related to the textile heritage of Łódź, a part of which was industrial design in the clothes industry and Polish fashion. Supporting such enterprises may help them to develop and select a group of leading urban events which would attract more tourists. In this case, we can observe a strong connection between these events and business tourism. In recent years, this kind of tourism has been the most rapidly developing tourism sector in Łódź.

Business tourism is currently a strategic type of activity in Łódź tourism sector. Due to its central location, the city has become a perfect place for company meetings, briefings of regional executives or business courses and conferences. Moreover, in the last decade, many international industrial corporations (e.g. Dell, Indesit, Bosch, Gillette) and service companies (e.g. Infosys) have made investments in the city. These firms also have an influence on the intensity of business activity. For instance, in 2014, Łódź hosted 1230 various business events, including 646 conferences, 402 corporation congresses, 151 motivational meetings and 31 fairs. All those events were attended by over 185 000 people (www.convention.lodz.pl/en/).

Furthermore, the city is located close to Warsaw—the capital of Poland. So far, it has been seen as a real curse on Łódź, because the more interesting investments were made in Warsaw as the more attractive city. However, at present, the situation seems to be changing. Firstly, Warsaw is the most expensive Polish city today, where the costs of living and investing are comparable with largest cities in Western Europe. On the other hand, Łódź is going through wide-ranging infrastructural changes. Moreover, a fragment of a new motorway has been open between Łódź and Warsaw, and soon the modernisation of fast railway joining the centre of Łódź with the centre of Warsaw is going to be finished. All this may encourage more firms to choose Łódź as the place for new investments, because of the close proximity to Warsaw and much lower costs. On the other hand, we may imagine a situation when Łódź starts to attract commuters who will live in Łódź and work in Warsaw.

How can this influence the Łódź festival sector? It must be stressed that it will be a noticeable and positive influence. The development of business tourism may have a positive effect on specialist events, such as Fashion Week or Łódź Design Festival, which are currently considered by the city authorities to be the most important events in the city. The business service centre created in Łódź may contribute to a bigger interest in this kind of events and enhance their development. Moreover, the potential inflow of businesspeople and immigrants to the city may create a larger and more affluent market for urban art festivals. These people will certainly become a group of well-off clients looking for entertainment and cultural and artistic experiences. They may obviously make up a significant group of consumers for many Łódź festival products.

It seems that in order for Łódź events to use this opportunity, several strategic steps should be taken. Firstly, a group of festivals should be chosen which should be supported financially and organisationally by the city authorities. This policy has already been implemented, but it seems that focusing on fashion or design events alone is not enough. Also some of the most valuable art festivals and those most strongly related to the cultural heritage of the city, such as the Łódź of Four Cultures Festival, should receive more substantial support. Apart from that, it would be a good idea to develop a cohesive common development strategy for Łódź festivals. The main role should be played by local and regional authorities, helped by private partners, i.e., the sponsors and patrons of these events. It would be a good idea to open one office organising events, to coordinate promotion and the purchase of services for strategic events. In this way, organisation would be coordinated and the cost of promotion and purchase of services would be lower.

The influence of Łódź festivals on the image space of the city is extremely important. Firstly, festivals change the way the city is perceived by its inhabitants and the groups from outside, e.g. tourists or potential immigrants and investors. A locality where many interesting events are held usually has a positive image. It is perceived as interesting, lively, having rich cultural heritage and innovative. Therefore, it brings positive connotations and is more often chosen as a tourist destination, a good place for a new investment or a place of potential immigration (see Rofe and Woosnam 2015). Łódź festivals certainly have a huge and already partly used promotional potential. It is a city where many festivals related to design, photography, graphics and fashion are held, which may be helpful in the creation of the Creative Łódź brand. Moreover, a lot of events represent art and high culture, e.g. film, theatre or music festivals. There is also (though in a changed formula) an interesting festival referring to the historical, multinational heritage of the city (Łódź of Four Cultures Festival), potentially still very valuable for the promotion of Łódź abroad. Generally speaking, such events have a positive effect on the image of the city and can be successfully used in the strategy of territorial marketing. They are a good base for creating the brand and image of a city with rich culture and multidimensional cultural heritage.

It must be stressed that adequate support, organisation and use of festival events may bring larger promotional benefits to Łódź. It becomes particularly important if we consider the fact that the city is still underrated in Poland, while abroad it is practically unknown. Unfortunately, all efforts to promote Łódź as a city of rich culture, history and as a festival city may be ruined by conflicts created around events such as Camerimage or the Dialogue of Four Cultures Festival.

Study tips:

- Explain what post-socialist cities are.
- Present the location and history of Łódź.
- Describe the causes and effects of the socio-economic transformations in Łódź after 1989.
- Present today's structure of Łódź festivals.
- Describe the effects of festival development for Łódź after 1989.

References

Adamus J, Paluch M (2012) Wydarzenia kulturalne w mieście poprzemysłowym. Przykład Łodzi. Turystyka Kulturowa 9:35–54

Buczkowska K (2009) Kulturowa turystyka eventowa, in: Buczkowska K, Rohrscheidt AM Współczesne formy turystyki kulturowej, Akademia Wychowania Fizycznego im. Eugeniusza Piaseckiego w Poznaniu, Poznań, pp. 91–118

Cudny W (2006) Festiwal Dialogu Czterech Kultur jako przykład strategicznego produktu turystycznego dla Łodzi. Turystyka i Hotelarstwo 10:117–128

Cudny W (2007) The role of the dialogue of four cultures festival in preserving the multinational and multicultural heritage of Lodz. In: Conference Proceedings Journeys of Expression VI: Diaspora Community Festivals and Tourism, Published by the Centre for Tourism & Cultural Change Leeds Metropolitan University United Kingdom

Cudny W (2011) Film festivals in Łódź as a main component of urban cultural tourism. Bull Geogr Socio-econ Ser 15:131–141

Cudny W (2012) Socio—economic changes in Lodz—results of twenty years of system transformation. Geografický časopis 64(1):3–27

Cudny W (2014) The phenomenon of festivals. Their origins, evolution and classifications, Anthropos 109(2):640–656

Cudny W, Rouba R (2011a) The role of Lodz festivals in promoting adventure tourism. Pol J Sport Tourism 18(4):265–274

Cudny W, Rouba R (2011b) Theatre and multicultural festivals in Lodz as a free time management factor among the inhabitants and tourists in the post-industrial city. Acta Geographica Universitatis Comenianae 55(1):3–22

Cudny W, Rouba R (2012a) Lodz Fashion Week as an example of a business event. Acta Geographica Universitatis Comenianae 56(1):45–58

Cudny W, Rouba R (2012b) Lodz—Poland. In: Cudny W, Michalski T, Rouba R (eds) Tourism and the transformation of large cities in the post—communist countries of Central and Eastern Europe. ŁTN, Wydawnictwo Uniwersytetu Łódzkiego, Łódź, pp 34–68

Cudny W, Rouba R (2012c) Festiwale muzyczne jako element funkcji kulturalno -rozrywkowej Łodzi. Zeszyty Naukowe Uniwersytetu Szczecińskiego, Ekonomiczne Problemy Usług 84 (699):593–606

Cudny W, Stanik E (2013) "Man in Danger" Media Festival in Lodz—the structure and motivations of the festival visitors. Annales UMCS Sectio B (Geographia, Geologia, Mineralogia et Petrographia) LXVIII(1):69–90

Cudny W, Ogórek P (2014) Segmentation and motivations of the attendees' of the Mediaschool Festival in Łódź, Poland. Bull Geogr Socio-econ Ser 24: 41–56

Cudny W, Gosik B, Rouba R (2011) Festiwal kultury chrześcijańskiej w Łodzi jako przykład produktu turystycznego—wydarzenie. Ekonomiczne Problemy Usług 66(648):89–100

Cudny W, Korec P, Rouba R (2012) Resident's perception of festivals—the case study of Lodz. Sociológia—Slovak. Sociological Review 44(6):704–728

Domański T (ed) (2008) Marka dla Łodzi. Wydział Studiów Międzynarodowychi Politologicznych. Uniwersytet Łódzki, Łódź, Strategiczne budowanie wizerunku miasta, Katedra Marketingu Międzynarodowego i Dystrybucji

Duvignaud J (1976) Festivals: a sociological approach. Cultures III(1):13–25

Dylik J, Dylik Z (1962) Nauka. In: Rosset E (ed) Łódź w latach 1945-1960. Towarzystwo Przyjaciół Łodzi, Łódź, pp 275–297

Dzieciuchowicz J (2009a) Ludność Łodzi do 1918 r. In: Liszewski S (ed) Łódź. Monografia miasta, Łódzkie Towarzystwo Naukowe, Łódź, pp 113–124

Dzieciuchowicz J (2009b) Ludność Łodzi od 1918 roku do lat 90. In: Liszewski S (ed) Łódź. Monografia miasta. Łódzkie Towarzystwo Naukowe, Łódź, pp 265–296

Evans G (2003) Hard-branding the cultural city—From Prado to Prada. Int J Urban Reg Res 27 (2):417–440

Garcia B (2004) Urban regeneration, arts programming and major events: Glasgow 1990, Sydney 2000 and Barcelona 2004. Int J Cult Policy (1):103–118

Gay P, Pryke M (eds) (2002) Cultural economy: cultural analysis and commercial life. Sage Publications Ltd., London

Jewtuchowicz A, Suliborski A (2009a) Życie społeczne i kulturalne Łodzi do 1918 roku. In: Liszewski S (ed) Łódź. Monografia miasta, Łódzkie Towarzystwo Naukowe, Łódź, pp 150–160

Jewtuchowicz A, Suliborski A (2009b) Życie społeczne i kulturalne Łodzi w latach 1918-1989. In: Liszewski S (ed) Łódź. Monografia miasta, Łódzkie Towarzystwo Naukowe, Łódź, pp 297–322
Jędrysiak T (2008) Turystyka kulturowa. Polskie Wydawnictwo Ekonomiczne, Warszawa
Kamiński A (1962) Łódź kulturalna—uwagi wstępne. In: Rosset E (ed) Łódź w latach 1945-1960. Towarzystwo Przyjaciół Łodzi, Łódź, pp 267–274
Klima E, Rosińska—Bukowska M (2012) Brand as an import ant element of regional development strategy at the example of Łódź. In: Michalski, T., Radchenko, A., Global, continental, national and regional conditions of local development. University of Gdańsk, National Academy of Public Administration at the President of Ukraine, Association of Doctors of Sciences on Public Administrations, Gdańsk-Kharkiv, pp. 90–98
Koter M (1969) Geneza układu przestrzennego Łodzi przemysłowej. PWN, Warszawa
Koter M (2009) Kształtowanie się tkanki miejskiej Łodzi do 1918 r. In: Liszewski S (ed) Łódź. Monografia miasta, Łódzkie Towarzystwo Naukowe, Łódź, pp 62–112
Koter M, Kulesza M, Puś W, Pytlas S (2005) Wpływ wielonarodowego dziedzictwa kulturowego Łodzi na współczesne oblicze miasta. Wydawnictwo Uniwersytetu Łódzkiego, Łódź
Liszewski S (2001) Model przemian miasta postsocjalistycznego. In: Jażdżewska I (ed.) Miasto postsocjalistyczne—organizacja przestrzeni miejskiej i jej przemiany cz. 2, XIV Konwersatorium Wiedzy o Mieście. Łódzkie Towarzystwo Naukowe, Łódź, pp 303–309
Liszewski S (2009) Wstęp. In: Liszewski S (ed) Łódź. Monografia miasta, Łódzkie Towarzystwo Naukowe, Łódź, pp 9–24
Marcińczak S, Sagan I (2011) The socio-spatial restructuring of Łódź, Poland. Urban Stud 48 (9):1789–1809
Matlovič R, Ira V, Sýkora L, Szczyrba Z (2001) Procesy transformacyjne struktury przestrzennej miast postkomunistycznych (na przykładzie Pragi, Bratysławy, Ołomuńca oraz Preszowa). In: Jażdżewska I (ed) Miasto postsocjalistyczne—organizacja przestrzeni miejskiej i jej przemiany (2). Wydawnictwo UŁ, Łódź, pp 9–21
Nartonowicz-Kot M (2009) Zarys dziejów Łodzi w latach 1918–1989. In: Liszewski S (ed) Łódź. Monografia miasta. Łódzkie Towarzystwo Naukowe, Łódź, pp 169-264
Ondoš S, Korec P (2008) The rediscovered city: a case study of post-socialist Bratislava. Geografický časopis 60(2):199–213
Pawlusiński R (2007) Wydarzenia kulturalne a rozwój turystyki w Krakowie, In: Stasiak A (ed) Kultura i turystyka razem czy oddzielnie. Wydawnictwo WSTH w Łodzi, Łódź, pp. 219–235
Puś W (1987) Dzieje Łodzi przemysłowej (zarys historii). Muzeum Historii Miasta Łodzi, Centrum Informacji Kulturalnej, Łódź
Rofe MW, Woosnam CL (2015) Festivals as a vehicle for place promotion: cars, contestation and the creative city ethos, Landscape Research, article published online. doi:10.1080/01426397.2015.1078457
Ray R, Sayer A (eds) (1999) Culture and economy after the cultural turn. Sage Publications Ltd., London
Scott AJ (2010) Cultural economy and the creative field of the city. Geografiska Ann Ser B, Hum Geogr 92(2):115–130
Skibińska M (2012) Szanse i zagrożenia łódzkich festiwali w opinii ich organizatorów. Uniwersytet Łódzki Filia w Tomaszowie Mazowieckim, supervised by Dr Waldemar Cudny
Słodczyk J (2001) Przestrzeń miasta i jej przeobrażenia. Wydawnictwo Uniwersytetu Opolskiego, Opole
Stanisławska A (2007) Imprezy kulturalno-rozrywkowe jako element identyfikacji turystycznej Łodzi. In: Stasiak A (ed) Kultura i turystyka razem czy oddzielnie. Wydawnictwo WSTH w Łodzi, Łódź, pp 205–218
Sýkora L (2009) Post-socialist cities. In: Kitchin R, Thrift N (eds) International Encyclopedia of Human Geography. Elsevier, Oxford, pp 387–395
Sýkora L, Bouzarovski S (2012) Multiple transformations: conceptualising postcommunist urban transition. Urban Studies 49:43–60

Wieczorkowska J, Goździk J (2002) Rzeźba terenu: plansza VIII. In: Liszewski S (ed) Atlas Miasta Łodzi. Urząd Miasta Łodzi, Łódzkie Towarzystwo Naukowe, Łódź

Young C, Kaczmarek S (2008) The socialist past and postsocialist urban identity in Central and Eastern Europe. The Case of Łódź, Poland. Eur Urban Reg Stud 15:53–70

http://www.convention.lodz.pl/en/—website of the Łódź Convention Bureau. Accessed 20 May 2015

http://www.gddkia.gov.pl/—website of the General Management of Roads and Motorways in Poland. Accessed 22 May 2015

Chapter 6
The Role of Festivals in the Development of Contemporary Urban Space—A Model

Abstract This chapter is the last factual chapter in this book, which consists of two main parts. The first one presents the most important findings from the analysis of festivals, their development and influence on cities discussed in the book. The second part presents a model illustrating the development of festivals and their role in the urban space. The first section of this chapter refers to the earlier parts of the publication. It includes the most important conclusions from the work, also as regards its objectives formulated in the Introduction. In this way, the first part consolidates the findings regarding the concept and history of festival events. The main factors of festivals development are considered here. These are the factors which led to the festival boom in the twentieth and twenty-first centuries. This part also refers to the main research approaches and methods used in festival studies. The conclusions from the chapter concerning urban space festivalisation, including those concerning the impacts of events on the multifaceted urban space, are also presented. The reader would also find references to the case study of Łódź—a Polish post-socialist and post-industrial city here. In the second part of this chapter, there is a model of the development of festivals and their influence on urban space presented. It shows both the main festival development factors in recent decades and their influence on space. The model is universal enough to be used for an analysis of other cities of different sizes and geographical locations.

Keywords Synthesis of festivals impacts · Urban space · Scientific models · Development and impacts of festivals—a model · Festivals in post-socialist city

Summary

This book answers the questions concerning the role of festivals in urban development. The growth of festivals and their various impacts on the society, economy and space was an object of interest in many scientific disciplines in the twentieth century. At the turn of the 1980s and 1990s, a separate subdiscipline emerged, called event studies. It investigates the issues of the organisation and socio-economic impacts of various events, including festivals (see: Getz 2012). This publication, however, has a different aim than books or articles about festivals

W. Cudny, *Festivalisation of Urban Spaces*, Springer Geography,
DOI 10.1007/978-3-319-31997-1_6

published as a part of event studies or social sciences, such as sociology or anthropology. It is a publication of a different kind, which refers to numerous aspects of festivals and urban space. Intentionally, this book is devoted strictly to the problems of urban festivals, their development and impacts on urbanised spaces. For this purpose, the conception of the multidimensional geographical space is used (for details see among others: Lisowski 2003; Thrift 2003) and adapted to the description of urban space.

The main aim was not only to specify the concept of the festival (which was not at all simple due to the multitude of its definitions and conceptions—see among others: Falassi 1987; Janiskee 1980; Piette 1992; Quinn 2005), but also its impacts on cities, often referred to as festivalisation. Chapter 2 presented the main features of festivals as well as definition of the phenomenon, based on the review of the most important scientific and encyclopaedic definitions. The main features of the festival included the fact that it is an organised and unusual public event, combined with a competition or review; it contains elements of celebration and fun and it has a leading theme. As for its influence on the surroundings, the festival performs a number of social functions (e.g. it solidifies the identity and enlarges social capital, strengthens interpersonal relationships), modifies and develops culture (from local through regional to supra-regional), and is an element of the cultural heritage.

One of the parts of the book is devoted to the history of festivals, analysed from antiquity to contemporary times. In this part of the book most important contemporary factors of festival development were also presented. First of all the process of culture development, including popular culture, as well as greater scientific interest in culture, known in humanities as the cultural turn (see: Jacobs and Weiss-Hnrahan 2005; Radclife 2006; Hawkins 2013). The significance of culture increased in the second half of the twentieth century, first in the developed countries, due to the decline of industry (deindustrialisation), the development of the service sector and the increasing value of unique experiences in human life. These unique experiences became a value sought by members of experience societies, which formed in the West, in the twentieth century. The term was coined by Schulze (1992) and perfectly describes the process of searching for strong emotions, typical of rich societies. It is a result of the civilisational development, higher earnings, a larger amount of free time, spatial mobility, education or access to modern information technology. As a consequence, it led to a situation where culture and festivals belonging to the culture sector and providing such interesting and unique experiences became an important element of contemporary societies' lives, also economically (Scott 2010). The last issue discussed in Chap. 2 was the themes of different types of festivals. Referring to various scientific publications and own observations, the author identified several criteria for festival classification: attitude to religion, festival location, social class structure, division of authority and social roles, important moments in personal life, the season, the scale and rank, repeatability of events, the form of organisation and financing, the structure of festival guests and visitors, the theme, the historical situation and the geographical situation of special events.

An important scientific aim of this publication was also to describe the research approaches taken in various scientific disciplines concerned with festivals, e.g. in event studies or geography. From the didactic point of view, it was also important to specify the range of basic research methods used for analysing various aspects of festivalisation. All these issues were presented in Chap. 3. The main scientific approaches used in festival analysis conducted as a part of event studies were the studies of the impacts of festivals on culture, society and economy, as well as the analysis of festival management. Geographical studies included cultural, social and economic analysis, time-space analysis, as well as political, theoretical, general and historical analysis (see also: Cudny 2014). Chapter 3 also contained a description of the main research methods used in the scientific analysis of festival events: literature or Internet resources analysis, query, observation, surveys and questionnaire interviews, as well as the study of the economic and environmental effects of organising festivals. Moreover, the basic cartographic and statistical analysis methods, as well as the descriptive, model, monographic, SWOT and BCG techniques used for studying a festival or a group of festivals, were characterised.

In Chap. 4, the next aims of the book were achieved—this chapter presented the basic concepts and definitions related to festivalisation as well as presented festivalisation spaces and their classification. First, there is a definition of festivalisation, which encompasses the development of festivals and their impacts on the surroundings. Next, the main impacts exerted by festivals on space, as well as the concept of festival tourism, were described. The influence of festivals on urban space was described with reference to physical space, the space of socio-economic connections (unblocking space), image space, place space or ecological and cultural space. The most important impacts of festivals on urban space were divided into tangible, intangible, positive and negative. The main effects of urban festivals included creating a place for celebration and promotion of culture, and a space of pursuing human interests. From the social point of view, it is important that a space of social contacts, socialisation and social capital development is created. From the economic point of view, it is important that new jobs are created in festival organisation and in the services provided for festival tourists. Another significant element is the creation of a positive image of a given city—an element of territorial marketing as well as a positive living atmosphere (good living conditions) for its inhabitants. In the context of the creative class theory, we must not forget about a particular urban atmosphere created by festivals, attracting this group of people (see Florida 2002). On the other hand, we should remember that festivals may have a number of negative impacts (which might be called dysfunctions) affecting the space of socio-economic connections and image space, which has already been described in the literature on the subject (Gotham 2005; Waitt 2008).

Chapter 4 presented the most important types of festivalisation space. Festivalisation spaces encompass those elements of urban space which are most strongly influenced by festivals and in which their impacts are best visible. Festivalisation spaces were divided into measurable and immeasurable (according to their physicality), open and closed (according to their construction), urban and rural (according to localisation), permanent and temporal (according to the time they are used for the event purposes), public, private and hybrid (according to accessibility), as well as summer and winter (according to seasonality).

The next aim of the book was achieved in Chap. 5, which presented a detailed case study of the influence exerted by festivals on the multidimensional urban space. This part of the publication characterises the example of Łódź—a large Polish post-socialist and post-industrial city. The location is first described, later the history and socio-economic conditions of the city, as well as the transformations which took place in Łódź after the fall of communism in Poland in 1989. It must be stressed that the socio-economic changes in the country led to significant economic transformations in Łódź (deindustrialisation, the development of economy based on services; privatisation, an inflow of foreign investments, industry restructuring) as well as to important social changes (demographic slump, emigration). The transformations were similar to those which took place in the second half of the twentieth century in Western Europe, during the transition from fordist to post-Fordist economy. After this introduction, the Chap. 5 characterised the most important Łódź festivals, their history, themes and the inhabitants' of Łódź attitude to urban festivals. The reader can also find how festival organisers evaluated Łódź as a place to hold festivals.

As regards the effects of organising festivals, they include mainly the impacts concerning the subspace of socio-economic flows, the cultural and the image space. The events held in Łódź are visited mostly by the inhabitants of the city. They represent a wide range of themes, though many of them are typical cultural events, related to music, theatre or film. They offer the opportunity to pursue one's interests and participate in unique experiences; they also make it possible to develop urban culture and art. From the economic point of view, festivals do not play a major role in the city life. Naturally, at the time when the majority of them are organised, we can observe a certain inflow of festival tourists to Łódź, but it is not significant or long-lasting enough to have an influence on the urban economy. It should also be stressed that the development of festivals has been supported by the city authorities for years, in connection with a development strategy based on the cultural and entertainment function (culture-led regeneration strategy).

Łódź festivals certainly have a large promotional potential, which is already being partly used. There are many festivals related to design, photography, graphics and fashion, which may be helpful in creating the Creative Łódź brand. Łódź is currently being promoted as a creative city, which can be aided by this kind of events. They may also stimulate the further development of business tourism (which Łódź specialises in), connected with activities involving, for example, industrial design. Some of the events organised in the city represent art and high culture. There is also an interesting festival referring to the historical, multinational heritage of the city (The Łódź of Four Cultures Festival). Such events have a positive influence on the city image and may be successfully used in the strategy of territorial marketing. They may be the basis for creating the brand and image of a city with rich culture and multidimensional cultural heritage. As for the negative effects, we should mention several serious conflicts between the city authorities and the organisers of some events, which wasted a part of the positive influence on the image space in recent years.

Based on the analysis carried out in this book, it should be concluded that the initial research hypothesis (presented in the Introduction) has been verified positively. It was assumed that the second half of the twentieth century brought significant changes all over the world as regarded the approach to culture and its role in the space of contemporary cities (the cultural turn, creative and cultural economy). It was stated that culture is important for the social development, generating incomes for cities and regions, as well as creating new jobs. Many cities in the developed countries in Europe and in other parts of the world started to implement strategies based on the development of the cultural and culture-entertainment function. Strategies of this type started to be introduced in many post-industrial cities as well, where the economic (industrial) base had been shaken by the deindustrialisation process. It resulted in the development of the event and festival sectors. Those events were used for the development of urban economy and satisfied people's need for culture and entertainment, as well as generated festival tourism. In recent decades, festivals have been playing increasingly important social and economic functions in cities. A result of their intensive growth was the creation of a number of spaces where festivals function and which they influence. In the 1990s, similar processes started to take place in the European countries where communism had fallen. As a result, the processes of deindustrialisation, internationalisation of economy and globalisation appeared in post-socialist cities as well. In many of them, event—or festival-led regeneration strategies—started to be implemented. An example of such a city is Łódź—a Polish post-industrial and post-socialist city described in this work.

Model

Apart from the most important conclusions to the analysis conducted in this book, one of its main aims was to present a compact model illustrating the development of festivals and their role in the development of urban space. "A model is a system of assumptions made in a given science while analysing a problem, in order to make it simpler, easier or even possible to solve it. It is called a nominal model. A model can also be a system of objects, events or situations, analysed instead of a complicated or more difficult real system under study, but sufficiently similar to it (isomorphic) as regards its basic aspects. We call it a real model of a given system (Berezowski 1980, p. 36)". It can be said that no part of the surrounding universe is simple enough to be precisely reproduced. Therefore, in order to do that it is necessary to construct various types of reality models (see: Rosenblueth and Wiener 1945). Gilbert (1991, p. 73) is of a similar opinion, claiming that science rests on constructing models which represent given issues, problems and the relations among them. A model may be then a system of words, symbols, pictures, numbers, etc., which will make scientific communication possible and easier.

The first comprehensive models in human geography were created in the 1960s, when the earlier descriptive research approach was replaced with the analytical approach (see: Haggett 1965; Chorley and Haggett 1967). As regards urban studies, models illustrating the spatial structure of cities were already popular in the first half of the twentieth century. They were, e.g. the classical model of the land use structure, developed by Burgess in 1926 for Chicago, or Hoyt's sector model, or Harris and Ulman's multiple nuclei model of urban land use (see: Pacione 2009). Those models illustrated the distribution of various functions in space and explained them, using economic and sociological theories. Urban studies also make use of real models, illustrating the processes of functional-economic and social transformations. They can be found in works concerning the changes in post-socialist cities during the system transformation after the fall of communism (see among others: Cudny 2012; Sýkora and Bouzarovski 2012).

It was decided to present the development of festivals and their influence on the city space in the form of a concise graphic real model. It represents the process and factors of urban festivals development, in the form of a diagram. It also presents the most important impacts exerted by festivals on the following dimensions of urban space: physical space, the space of socio-economic linkages (unblocking space), image space, place space, ecological and cultural space (Fig. 6.1).

As shown by the model above, festivals have different origins and exert a wide range of impacts. They have been developing for centuries, and their role has been growing along with the rising level of human culture and civilisation. In the twentieth century, the civilisational development was so advanced that it produced perfect environment for the so called "festival boom". The factors which led to it include first of all the development of culture, art and entertainment, caused by civilisational changes such as an increase in the average earnings, the amount of

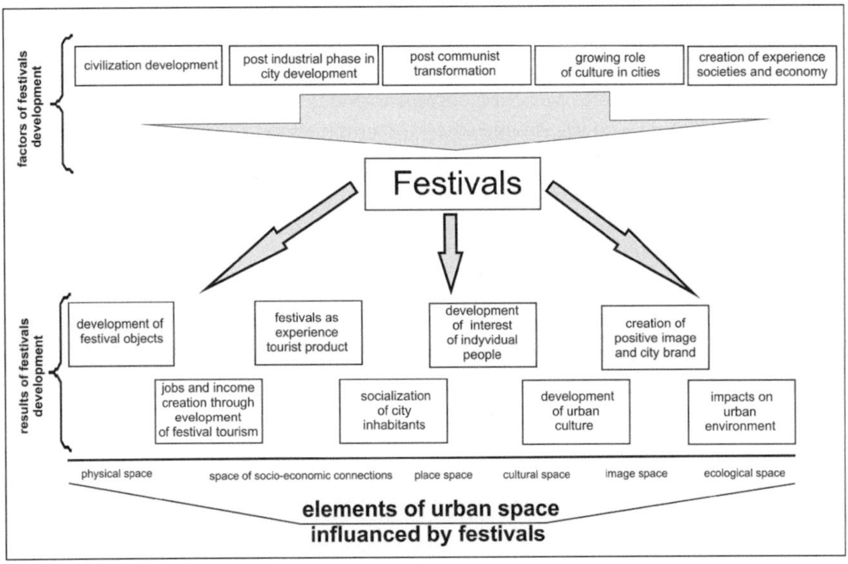

Fig. 6.1 A model illustrating the development of festivals and their role in the development of urban space. *Source* Author's compilation

free time and the level of education. That triggered a greater interest in culture, which became a common consumption good. It also started to be used as an economic development factor. Cultural products became a generator of profits and jobs (Fig. 6.1).

Those processes took place first in the developed countries, e.g. in Western Europe or North America, and later in the developing regions. A result of the transformations described above was the unique phenomenon of forming experience societies, i.e. societies consuming experience products. Such societies are usually highly developed and their members seek powerful and extraordinary impressions. These people have usually fulfilled their basic needs and make use of various services, including cultural and entertainment ones. Due to the tediousness of everyday life, they look for something new and unusual. One answer to the need of such experience products are festivals. They present various elements of human culture in an interesting and dynamic way. Attending a festival is often connected with travel as a part of festival tourism, which additionally increases the attractiveness of this kind of events.

The factors described above have led to the development of festivals on a mass scale, especially the urban festivals. They started to exert a strong influence on many dimensions of urban space (Fig. 6.1). Festivals became an element used for the economic stimulation of the cities which have lost or significantly decreased their industrial function as a result of deindustrialisation. Many cities supported the functioning of this sort of events, expecting a development of festival tourism and favourable effects connected with creating a positive image of the city of culture and art. Festivals play an important role in the inhabitants' socialisation (making new interpersonal contacts, the development of social capital), as well as enable the inhabitants and tourists to pursue their interests. They also influence the physical space by building festival facilities, as well as the natural environment (e.g. by increasing the level of pollution).

As we can see from the description above, festivals are nowadays a significant and permanent element of urban culture. They have become an element typical of the majority of contemporary urban centres. The long centuries of the civilisational development led to the popularisation of festivals, especially in the cities, as well as to their growing impact on people and the space in which they live. This complicated process of festivalisation was an object of analysis in this book.

References

Berezowski S (1980) Metody badań w geografii ekonomicznej. WSiP, Warszawa

Chorley RJ, Haggett P (ed) (1967) Socio-economic models in geography. London, Open University Setbook, Methuen & Co

Cudny W (2012) Socio-economic changes in Lodz—results of twenty years of system transformation. Geografický časopis 64(1):3–27

Cudny W (2014) Festivals as a subject for geographical research. Geografisk Tidsskrift-Danish J Geogr 114(2):132–142

Falassi A (1987) Festival: definition and morphology. In: Falassi A (ed) Time out of time. University of New Mexico Press, Albuquerque, pp 1–10
Florida R (2002) Bohemia and economic geography. J Econ Geogr 2(1):55–71
Getz D (2012) Event studies. Theory, research and policy for planned events. Routledge, London and New York
Gilbert SW (1991) Model building and a definition of science. J Res Sci Teach 28(1):73–79
Gotham KF (2005) Theorizing urban spectacles. Festivals, tourism and the transformation of urban space. City 9(2):225–246
Haggett P (1965) Location analysis in human geography. Edward Arnold, London
Hawkins H (2013) Geography and art. An expanding field: Site, the body and practice. Prog Hum Geogr 37(1):52–71
Jacobs MD, Weiss-Hanrahan N (2005) The Blackwell companion to the sociology of culture. Blackwell Publishing Ltd., Malden
Janiskee B (1980) South Carolina's harvest festivals: rural delights for day tripping urbanites. J of Cult Geogr 1(1):96–104
Lisowski A (2003) Koncepcje przestrzeni w geografii człowieka. Uniwersytet Warszawski, Warszawa
Pacione M (2009) Urban geography: a global perspective. Routledge, London–New York
Piette A (1992) Play, reality, and fiction: toward a theoretical and methodological approach to the festival framework. Qual Sociol 15(1):37–52
Quinn B (2005) Art festivals and the city. Urban Res 42:927–943
Radclife SA (2006) Culture in development thinking: geographies, actors, and paradigms. In: Radclife SA (ed) Culture and development in a globalizing world: geographies, actors and paradigms. Routledge, London, pp 1–29
Rosenblueth A, Wiener N (1945) The role of models in science. Philos Sci 12(4):316–321
Schulze G (1992) Die erlebnis gesellschaft. Kultursoziologie der gegenwart, Campus, Frankfurt
Scott AJ (2010) Cultural economy and the creative field of the city. Geogr Ann Ser B Hum Geogr 92(2):115–130
Sýkora L, Bouzarovski S (2012) Multiple transformations: conceptualising postcommunist urban transition. Urban Stud 49:43–60
Thrift N (2003) Space: the fundamental stuff of geography. In: Rice S, Valentine G (eds) Key concepts in geography Holloway SL. Sage, London, pp 95–108
Waitt G (2008) Urban festivals: geographies of hype, helplessness and hope. Geogr Compass 2(2):513–537

MIX
Papier aus verantwortungsvollen Quellen
Paper from responsible sources
FSC® C105338

If you have any concerns about our products,
you can contact us on
ProductSafety@springernature.com

In case Publisher is established outside the EU,
the EU authorized representative is:
Springer Nature Customer Service Center GmbH
Europaplatz 3, 69115 Heidelberg, Germany

Printed by Libri Plureos GmbH
in Hamburg, Germany